together at the table

RURAL STUDIES SERIES

Leif Jensen, General Editor
Diane K. McLaughlin and Carolyn E. Sachs, Deputy Editors

The Estuary's Gift:
An Atlantic Coast Cultural Biography
David Griffith

Sociology in Government:
The Galpin-Taylor Years in the U.S. Department of Agriculture, 1919–1953
Olaf F. Larson and Julie N. Zimmerman
Assisted by Edward O. Moe

Challenges for Rural America in the Twenty-first Century
Edited by David L. Brown and Louis Swanson

A Taste of the Country:
A Collection of Calvin Beale's Writings
Peter A. Morrison

Farming for Us All:
Practical Agriculture and the Cultivation of Sustainability
Michael Mayerfeld Bell

together at the table

Sustainability and Sustenance in the American Agrifood System

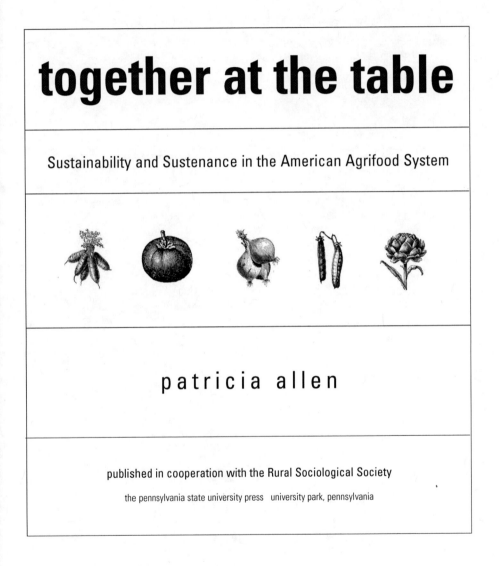

p a t r i c i a   a l l e n

published in cooperation with the Rural Sociological Society

the pennsylvania state university press   university park, pennsylvania

LIBRARY OF CONGRESS CATALOGING-IN-PUBLICATION DATA

Allen, Patricia, 1954–
  Together at the table : sustainability and sustenance in the
  American agrifood system / Patricia Allen.
      p.   cm. — (Rural studies series of the Rural Sociological Society)
  Includes bibliographical references and index.
  ISBN 978-0-271-02473-8 (cloth: alk. paper)
  ISBN 978-0-271-02977-1 (pbk: alk. paper)
  1. Food industry and trade—United States.
  2. Agricultural industries—Environmental aspects—United States.
  3. Sustainable agriculture—United States—Economic aspects.
  4. Natural foods industry—United States.
  I. Title.   II. Series.

  HD9005 .A69 2004
  338.1'0973—dc22
  2004010245

contents

acronyms

AFI	alternative food institution
AFL-CIO	American Federation of Labor and Congress of Industrial Organizations
ALRA	Agricultural Labor Relations Act
BIFS	Biologically Integrated Farming Systems
BIOS	Biologically Integrated Orchard Systems
CalSAWG	California Sustainable Agriculture Working Group
CANFIT	California Adolescent Nutrition and Fitness
CASA	California Alliance for Sustainable Agriculture
CFP	Community Food Project
CFS	community food security
CFSC	Community Food Security Coalition
CSA	community-supported agriculture
CSARE	Consortium on Sustainable Agriculture Research and Education
DBCP	A banned pesticide
DDT	A banned pesticide
EBT	Electronic Benefit Transfer
EWG	Environmental Working Group
FLO	Fair Trade Labeling Organizations
FMNP	Farmers' Market Nutrition Program
FPC	food policy council
GAO	U.S. General Accounting Office
IFAFS	Initiative for Future Agriculture and Food Systems
IFFS	Integrated Food and Farming Systems
IFOAM	International Federation of Organic Agriculture Movements
IFS	Integrated Farming Systems
LISA	Low-Input Sustainable Agriculture program
MAI	Multilateral Agreement on Investment

NAACP	National Association for the Advancement of Colored People
NAFTA	North American Free Trade Agreement
NGO	nongovernmental organization
NIMBY	"not in my backyard"
NRC	National Research Council
OECD	Organization for Economic Cooperation and Development
PAC	political action committee
SAI	Social Accountability International
SAN	Sustainable Agriculture Network
SARE	Sustainable Agriculture Research and Education
SAREP	Sustainable Agriculture Research and Education Program
SASA	Social Accountability in Sustainable Agriculture
SFMNP	Seniors Farmer's Market Nutrition Program
UC SAREP	The University of California's Sustainable Agriculture Research and Education Program
UCLA	University of California, Los Angeles
UFW	United Farm Workers
USAID	U.S. Agency for International Development
USAS	United Students Against Sweatshops
USDA	U.S. Department of Agriculture
WIC	Special Supplemental Nutrition Program for Women, Infants and Children
WTO	World Trade Organization

To those who have the courage and imagination to work for justice.

acknowledgments

During the years this book was in the making, many people provided support and inspiration.

Ironically, perhaps, my greatest inspiration came from a child. In her own inimitable way, my daughter, Kitt, reminds me daily that we can remake the world into a better place. She has a wonderful sense of perspective and keeps me laughing. My partner, David, has been unreasonably supportive for lo, these many years. His beneficence ranges from uncomplainingly enduring my work schedule to doing the dishes. Without his generosity of spirit this book never would have seen the light of day.

Dennis Taku gave his time and talent to teach my daughter to play baseball. Watching her excel as a pitcher and batter in competitive baseball has been a singular joy. At least on the baseball fields of Santa Cruz another gender barrier is crumbling.

My cat, Baby, provided the perspective and calm assurance that only cats can. However, he once he licked all of the fur off his arms in a show of neurotic solidarity during a particularly stressful time for me. He died at 21 1/2 years old during the completion of the book.

My family both diminished and grew during this time. Both my mother and my father died during the writing of this book. In dealing with the horror that was their deaths, however, I found a true friend in my brother, Tim. Deb Van Dusen, Sandra Meucci, Beth Hill, Lyn Garling, and Christina Cecchettini know how important they have been in my family.

Many people played a part in developing this book. Jim O'Connor taught me to always look under the surface and helped me work through many of my initial conceptualizations. Margaret FitzSimmons and Carolyn Sachs talked through many of my thoughts with me and provided kind and strong encouragement. Fred Buttel and Andy Szasz helped me work through ideas and put pieces together. I have also benefited from the opportunity to discuss agrifood ideas and issues in the Agrifood Seminar at UC Santa Cruz with luminaries such as Bill Friedland, David Goodman, Melanie DuPuis, and Julie Guthman.

Carol Shennan, Director of the Center for Agroecology and Sustainable Food Systems, provided precious material and structural support. Without her help, this book would still be in a pile on my desk.

Many people helped with the research on this book and my other projects. At different times Richard Rawles, Jan Perez, Hilary Melcarek, Valerie Kuletz, Marty Kovach, Phil Howard, Mike Goodman, Dave Carlson, and Martha Brown provided invaluable assistance.

I am indebted to Peter Potter and Leif Jensen at Penn State University Press for their professionalism and forbearance. Andrew Lewis, my copy-editor, is one of those rare individuals who can see both the forest and the tiniest details of the trees as well as their undergrowth.

One theme that comes up again and again in the research of the history of social movements is the crucial importance of key individuals. This was clearly the case in the development of alternative agrifood movements and the programs I discuss in this book. These are the people who have the vision and the leadership to engage people to build a better world. While the leaders in these movements are recognized, there are also countless others in less visible positions who enable the progress of social movements. I would like to acknowledge them, although I don't know them by name. You know who you are.

<table>
<tr><td>**1**</td><td rowspan="2">sustainability and sustenance in the
agrifood system</td></tr>
<tr><td>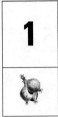</td></tr>
</table>

Everywhere you look these days there are signs that people are beginning to take charge of their food system. In a cafeteria in Los Angeles, children make their lunchtime choices at fresh-fruit and salad bars stocked with local produce. In a community garden in New York, low-income residents are producing organically grown fruits and vegetables for their own use and for sale. In Madison, Wisconsin, shoppers make their selections from a bounty of choices at a vibrant farmers' market. In universities across the country, faculty members research and students study organic farming. In San Francisco, "at-risk" teenagers run an organic food business. On a farm in Santa Cruz, California, unionized farmworkers grow and harvest organic strawberries. In Washington, D.C., legislators develop new policies and programs to promote sustainable agriculture and community food security. These kinds of activities span the entire United States, from Hawaii to Maine, as diverse groups of people work to construct alternatives to the conventional practices, discourses, and institutions of the contemporary agrifood system. In the United States much of this work has been spearheaded and encompassed by the movements for sustainable agriculture and community food security. The goals of these movements are to reconstruct the agrifood system to become more environmentally sound, economically viable, and socially just.

Alternative agrifood activities and actions are the result of both increased knowledge about the agrifood system and increased understanding that the system can be changed. Today's newspapers and newsrooms, the oracles of modern times, increasingly lead with stories about food and agriculture. Occurrences of mad cow disease, the mysterious infiltration of the food supply by genetically modified foods, pesticide drift near elementary schools, charity food distribution for working people, the transformation of farms into shopping centers, epidemic rates of obesity—all are regularly placed at the forefront of public consciousness. Every day in the United States resources are depleted, toxins enter the food chain, people go hungry, and the gap between the rich and the poor grows at an accelerating rate. Yet many people do not feel helpless in the face of this staggering array of environmental and social problems. They realize that, as the country moves further and further from democratic practice, these conditions have been accompanied and enabled by a process that wrests decision making away from ordinary people. They witness the failure of electoral politics and political parties to solve agrifood problems, a situation they fear can only get worse, as the decision-making ability of elected governments is superseded by the power of global capital to limit choice. They have decided that it is time to take matters into their own hands.

In many places and in many different ways people are struggling to improve conditions in the agrifood system. Not content to let food production, distribution, and quality be defined and determined by faceless others, they have taken action. Consciously or not, they are part of a new assemblage of movements sweeping the nation, movements for alternative food and agriculture. The issues with which these groups are concerned include food safety, access to food, environmental degradation, and rural development. Together they are addressing these basic issues of sustenance and sustainability—to reconfigure the agrifood system to meet people's food needs both for the present and for the future.

Two movements figure prominently in these efforts: a movement for sustainable agriculture and a movement for community food security. The concerns they address are closely related but have somewhat different emphases. The sustainable agriculture movement has focused primarily on production-centered issues, such as environmental degradation and the viability of the family farm. The community food security movement has centered more on issues of distribution and consumption, such as food access and nutrition problems. These movements are related in different but complementary ways, and the increasing consumer demand for pesticide-free, organic, non–genetically modified food has only strengthened the ties between them. Because the issues they address are so important, they have attracted a broad range of participants and have become significant social movements.

Social movements are efforts to change widespread existing conditions—political, economic, and cultural. The multiple strategies that social movements employ to achieve their objectives can be quite varied. Alternative agrifood movements in the United States operate primarily at two levels: at the level of developing alternative practices, such as those just described, and at the level of changing institutions. Historically, many social movements have chosen to operate outside the state, having little faith in the sociopolitical process and power structures that excluded their concerns in the first place. In America's agrifood system, for example, those who have been able to influence political decision making have been primarily producer groups and food industries little interested in issues of agricultural sustainability or food security. Yet because of the central role of government in the American agrifood system, the movements for sustainable agriculture and community food security have had to engage public institutions at local, state, and federal levels. Therefore, in addition to working on many other fronts, these alliances of farmers, environmentalists, consumers, and scientists have sought and achieved a "place at the table" in major food and agricultural institutions. Ideas that were once anathema, in the case of sustainable agriculture, or unknown, in the case of community food security, have become part of the policy, research, and education agendas of these institutions.

What is the effect of these efforts to create change in the agrifood system at both community and institutional levels? Although there has been no comprehensive evaluation of these efforts, it would seem that they have already begun to improve conditions of everyday life for those who have not been well served by the conventional agrifood system. For example, the creation of a farmers' market in an inner city where there was previously little or no access to fresh fruit and vegetables is surely a positive development. Similarly, providing institutional funding to teams of researchers working with farmers to develop environmentally sound farming practices is an important step toward resource conservation in agriculture. These incremental improvements, significant in themselves, also provide openings for catalyzing further changes as programs and networks expand. The people involved in these diverse efforts can coalesce into a powerful social movement for restructuring and transforming the agrifood system in the direction of greater environmental soundness and social justice.

Alternative agrifood movements may also possess significant potential to develop into even broader movements for social and environmental change. For example, the introduction of genetically modified organisms into the food supply has become a powerful catalyst for social activism, spanning issues of food safety, sustainability, equity, biodiversity, and democracy. Agricultural sustainability and food security are important to each and every person, regardless of economic or social class. Moreover, as discursive symbols, both sustainability and food security are enormously powerful. Youngberg and others (1993) suggest that in its emotional appeal and evocative meanings, sustainability is on par with concepts such as freedom, liberty, and democracy. Yet the extent to which alternative agrifood movements and their activities help create substantial change in the direction of greater environmental sustainability, social equity, and food security remains unclear. In other words, analysis of these rapidly developing alternative discourses and practices lags behind their proliferation in communities and institutions.

This book is a first step toward such an analysis. In it I explore the discourses and practices of alternative agrifood movements and actions and the translation of movement ideals into practice. I focus primarily on the sustainable agriculture and community food security aspects of the alternative agriculture movement. Specifically, I examine how the ideas and practices of sustainable agriculture and community food security have been woven into the dominant agrifood institutions in the United States In addition, I explore the possibilities this process may hold for improving social and environmental justice in the American agrifood system.

Social Movements and Social Change

Throughout human history, social change has been brought about by people organizing themselves to correct a perceived injustice or inequity. In the United States, food safety laws, women's suffrage, the abolition of slavery, workers' rights to unionize, antihunger programs, the end to the Vietnam War, our very independence as a nation—all were brought about by the collective actions of ordinary people.

There has been some debate about whether alternative agrifood efforts like sustainable agriculture or community food security actually represent social movements at all, or whether they behave more like something more modest, such as special interest groups or affinity groups. This raises the question: What is a social movement? While social scientists devote much thought and analysis to the definition of social movements, Cohen (1985) has pointed out that there is little agreement among theorists on what a social movement is exactly and how it differs from a political party or interest group. Assigning the term "social movement" to a group of actors there-fore remains somewhat arbitrary. Many different phenomena have been categorized as social movements, including public-interest lobbies, reli-gious movements, revolutions, and political reform movements (McAdam et al. 1988). The term generally refers to persistent, patterned, and widely distributed collective challenges to the status quo. Collective action becomes a movement when participants refuse to accept the boundaries of established institutional rules and routinized roles. For Darnovsky and oth-ers (1995), social movements are collective efforts by socially and politi-cally subordinated people to challenge the conditions and assumptions of their lives.

Within this framework, can any alternative agrifood effort legitimately be called a social movement? To answer this question, I refer to Scott (1990), who proposes that a social movement is a collective actor consti-tuted by individuals who understand themselves to have common interests and a common identity. The issue of self-perception is crucial to this def-inition. That is, if the participants in sustainable agriculture and commu-nity food security groups refer to what they are doing as a social movement—and they do—there is little purpose in scholarly questioning of their terminology. However imperfectly articulated and integrated, a large group of people working together to achieve sustainability and com-munity food security is considered to be, and should be referred to as, a social movement.

Alternative agrifood movements have similarities in themes and strate-gies with other progressive social movements. Merchant (1992) situates the movement for sustainable agriculture within the environmental and ecofem-inist movements. These types of movements, which began to take shape in the 1970s, are new in the sense that their objectives are not delimited by objectives such as increased workers' power or national liberation, as were "old" social movements.[1] There is nothing new about concerns like women's rights, peace, and the environment. These issues have long been with us, but were probably suppressed in the old social movements (Frank and Fuentes 1990). Common themes of new social movements are strug-gles for a democratic, postpatriarchal society (Cohen 1985), often centered on specific political goals or recognition of rights. New social movements are increasing in strength and importance; they inspire and mobilize people more than the "old" ones do (Frank and Fuentes 1990). These movements

1. To bypass the issue of new versus old Cohen (1985) has suggested the term "contem-porary" social movements to describe the movements that have developed since the 1970s.

are driven not only by abstract social issues but also by concerns about their participants' own life conditions and identities, issues that they experience in daily life. Perhaps because of this immediacy, these movements have become quite powerful.

Discourse and Social Movements

In this book I focus on discourse because of its centrality in the constitution and efficacy of social movements. By "discourse" I mean the ensemble of social, political, and cultural languages, meanings, codes, and relationships that construct, maintain, or challenge the social order. It is the process through which social reality comes into being.

Discourse is what forms and maintains social movement identity. In fact, for some, discourse is primarily what a social movement *is*. For Eyerman and Jamison (1991: 3), for example, the concepts, ideas, and intellectual activities—the cognitive praxis—of a social movement are what give the movement its identity and its particular meaning. For them, cognitive praxis is the core activity of a social movement, and this cognitive territory is what transforms a group of individuals into a social movement. "It is precisely in the creation, articulation, and formulation of new thoughts and ideas— new knowledge—that a social movement defines itself in society" (Eyerman and Jamison 1991: 3). Discourse is not only constitutive of social movements; it is also one of the primary tools movements employ to work toward social change.

For many analysts, the primary power of social movements is discursive, that is, it lies substantially in their ability to challenge dominant perspectives and priorities by raising new issues, changing popular consciousness, and opening new arenas of public policy. Power is embodied in and exercised through discourse. Control of discourse by institutional and societal power holders is a key factor in maintaining power (Fairclough 2001). The discursive construction of reality is a crucial realm of power for social movements that do not control major economic resources or the formal political process. While government and economic resources are major loci of power in society, another is ability to define situations (Wallerstein 1990). Discursive struggles are therefore crucial arenas for instigating changes in cultural and material conditions and within institutions.

One of the key functions of a social movement is to challenge and "rehabilitate" social institutions, to "reform" public space so that new ideas and relationships can develop. It is through discourse that dominant ideas within organizations and institutions are produced, reproduced, contested, and transformed (Fairclough 1994: 10). The relationship between the discourse of social movements and that of social institutions is dialectical. That is, as movements reshape institutions, institutions also reshape movements. Social institutions both determine and are produced by discourse. Discourse simultaneously reflects and creates social reality. It is in this discursive space that the present study is located.

Studying Alternative Agrifood Movements

Since most alternative agrifood ideas and practices have emerged relatively recently (or only recently come under academic scrutiny), research analyzing alternative agrifood discourses and practices is still in its infancy. According to Kloppenburg and others (1996) this relative paucity of research on alternatives to the agrifood system is also related to the fact that analysts of the food system have tended to focus more on the problems of agribusiness, and less on the work being done to solve those problems. Increasingly, however, scholars are looking closely at the development of these alternatives; research to date on alternative agrifood practices focused mostly on one of three approaches (Allen et al. 2003)—identification, classification, and analysis.

The first approach consists primarily of identifying and describing these alternatives—a kind of affirmation that people are actively engaged in developing alternative food pathways and institutions (see, for example, Henderson 1998). The second has had a more instrumental focus, evaluating various types of agrifood alternatives in terms of their potential for helping different populations or sectors such as small-scale farmers, food-based entrepreneurs, or regional economies (e.g., Ilbery and Kneafsey 1999 and Kolodinsky and Pelch 1997). The third approach focuses on analyzing specific expressions of alternative agrifood efforts, such as direct marketing (e.g., Hinrichs 2000) or community-supported agriculture (CSA) (e.g., DeLind and Ferguson 1999). It still remains for researchers to study the constellation of agrifood alternatives. In an effort to develop this research agenda, I have undertaken in this book to analyze the discourse and practices of the alternative agrifood movement and their integration into traditional agrifood institutions in the United States.

As I have argued before, this kind of analysis is important for enabling alternative agrifood efforts to accomplish their goals and minimize potentially contradictory outcomes. Those working in alternative food movements have neither the time nor often the inclination to study the larger context of their work. While committed people work in many different areas of the food system to effect change, those embroiled in direct action, whether on farms, in nongovernmental organizations, in laboratories, or in agrifood businesses, rarely have the opportunity to analyze their efforts. Yet this type of analytical process can reveal possibilities for and obstacles to success that may be obscured by the demands of day-to-day work. Marsden and Arce (1995) point out that without close, empirical studies of food systems, we are likely to miss not only understanding how such systems work but also—and perhaps more important—how they might *change*.

This work also attempts to fill a gap in the study of social movements. Eyerman and Jamison (1991) write that sociologists have generally ignored the cognitive dimensions of activities in the movements they study, focusing instead on actions such as the mobilization of resources, organizational methods, and campaign strategies. For many sociologists knowledge and identity are seen as nonempirical objects and therefore outside the range of what can be studied. Other scholars of social movements focus on the identities of the movements, but study them primarily by reference to theories of social change and philosophies of history.

My subject in this book is primarily the discourse of the alternative agrifood movements in the United States generally and in California in particular. This subject matter includes the assumptions shared by participants in the movement as well as the specific topics or issues around which the movements are created, that is to say, their cognitive content. What are the core assumptions and positions of the movements? How far do they take us on a path to an environmentally sounder and more equitable agrifood system? I am also interested in how alternative agrifood discourses have been integrated into major agrifood institutions. What has been the record and effect of this integration? What is the potential of the alternative discourses and practices supported by the movements themselves?

The data for this analysis come from several sources. These include the projects funded by public programs in sustainable agriculture and community food security, publications by leaders and participants in the alternative agrifood movement, interviews with key people in these traditional agrifood institutions and alternative agrifood organizations, surveys and interviews of farmers and consumers, and my own observations as a long-time participant in alternative agrifood movements.

I also used textual sources: institutional grant programs in sustainable agriculture and community food security; published documents, including program reports, pamphlets, and manuscripts written by program leaders; and alternative agrifood movement publications, presentations, and conference programs. Institutional grant programs in sustainable agriculture and community food security are social forms where discourse and practices are evident and formalized. In these we can see which ideas and practices are preferred and privileged, and which are downplayed or omitted. Published documents written by program leaders reveal collective institutional priorities and perspectives. Alternative agrifood movement publications, presentations, and conference programs represent the self-identified perspectives and priorities of alternative agrifood movements.

Of course, real people carry out relationships between and within institutions. An investigation of people's self-understanding is crucial to learning more about the meaning and potential of sustainable agriculture and community food security discourse and practice. Therefore, I interviewed key people in the movements and the institutions. These interviews are "triangulated" by my own observations at alternative agrifood conferences and meetings, based upon my "position" within the movements. Because I have myself been involved in the movement for sustainable agriculture for nearly twenty years, and in the community food security movement almost since its inception, I am also a participant observer. I initiated and organized the first University of California conference on sustainable agriculture in January 1985, at a time when the very concept of sustainability was considered heretical within the agricultural establishment. In 1995 I organized a community food security project in Santa Cruz, California. I collaborated on developing the original proposal and was a participant in a California organizational collaboration on agricultural sustainability and food-system issues, and I have attended and taken part in numerous alternative agrifood meetings and activities over the years. Thus I have had many opportunities to bypass the academic isolationism that Epstein (1990: 39) criticizes in the study of social movements. In her view, the absence of a "vital intellectual connection" to social movements leads researchers to develop theories "more about than for the movements."

Discovering how people working in the alternative agrifood movement and agrifood institutions view the world and how they see their place in challenging and reshaping the agrifood system represents an essential step for better understanding the sites of and possibilities for change in the agrifood system. Yet these perspectives are rarely studied. According to Kloppenburg and others (2000), conceptual framings of alternative food systems have been devised primarily by academics and policy specialists, but so far, none of these perspectives reflects the full range of understandings among those producers and consumers who constitute the bulk of the movement. In their study of the meaning of food-system sustainability within a broad cross section of the alternative agrifood community, Kloppenburg and others found that popular meanings of sustainability often differed significantly from the definitions of academics and professional advocates. They assert that it is essential to include the perspectives of "ordinary people," who are, after all, the "principal agents of change in the efforts to recreate the food system."

An intensive study of subjectivity is beyond the scope of this book; however, I draw on three studies in which I was involved to include the perspectives and priorities of participants in alternative agrifood movements and practices. The earliest of these is a survey of California agrifood organizations that I conducted in the fall and winter of 1996–97 under the auspices of the California Alliance for Sustainable Agriculture (casa). This survey, the first of its kind, gathered information on organization mission statements, conceptions of sustainability and food security, and projects and activities.[2] The initial list of organizations was compiled by casa members and was supplemented through conversations with these initial groups. Organizations were targeted because they are more influential than individuals and because their perspectives are the products of larger discussions and deliberations and more closely represent the views of their constituencies. Although results from the survey are more illustrative than definitive, they nonetheless provide a picture of the perspectives and priorities of respondent organizations. We received 71 questionnaires out of 196, a response rate of 36 percent.[3]

The second study is one I conducted with a research team at the University of California, Santa Cruz, on alternative agrifood institutions (afis) in California. afis are the collective efforts of people to build food systems that are more environmentally sound and socially just than the conventional food system. This research focused on the subjectivity of "agents,"

---

2. Although a sustainable agriculture survey was conducted in California in 1988 (see Francis 1988), this survey was administered only to farmers and focused on characterizing farms using practices oriented toward sustainability.

3. Originally the questionnaire was sent to 238 organizations. A number of organizations (37) were subsequently removed from the list after we discovered that the organization no longer existed or was part of another organization. Another 5 organizations that returned questionnaires were removed from the list because they did not provide information requested or turned out not to be nonprofit organizations working in California.

that is to say, the people who actually do the work of developing agrifood alternatives in California. Our goal was to document how people express agency in reaction to the problems they perceive in the agrifood system as well as to reflect their self-perceptions of their actions. In the first phase of this research, we focused on the leaders, since leadership is considered to be a crucial ingredient in the trajectory and success of these organizations. Through her work with numerous community-based food organizations, Feenstra (1997) determined that the first key element for developing sustainable, equitable food systems is leadership by clearly identifiable leaders who can build strategic relationships. Thus, in conducting this research project, our hope was not only to gather information about an organization and its activities but also to learn more about the perspectives of the leaders who guide and direct each organization.

For this phase of the study we identified eighty California organizations that fit within a general typology of alternative agrifood organizations. Of these eighty organizations, we selected forty-five that represented a range of activities intended to change the way food is produced, consumed, or distributed. Programs offered by these organizations included alternative agrifood education programs, therapeutic agriculture programs, local and regional food labels, agrifood microenterprises, urban agriculture and community gardens, food policy advocacy, farm-to-school programs, community-supported agriculture, and farmers' markets. Contacts with these organizations resulted in a list of thirty-seven that were still in existence and able to participate in the study. Geographically, the distribution of our study sample reflected the population densities of these alternative agrifood organizations in California. Organizations were often located in both northern California (mostly near the San Francisco Bay area) and southern California (mostly in and around the Los Angeles area). Our goal in this study was neither statistical rigor nor generalizability. Rather, it was to learn about the worldviews and transformative potential of alternative food efforts by listening to the perspectives and insights of their leaders as expressed through in-depth interviews.

Research team members conducted semi-structured interviews with organization leaders, primarily face-to-face, supplemented by telephone interviews where in-person interviews were not possible. In each case, the interviewee was sent a list of the interview questions beforehand so they could provide thought-out, rather than spur-of-the-moment, responses. The questionnaire was designed to collect basic information about the organization's history, activities, obstacles, and influences. It also provided opportunities for AFI leaders to share their perceptions of key problems and solutions in the food system, their vision for a better food system, and their motivations for being involved in alternative food work. In these interviews we collected basic information about the organization's history, activities, obstacles, and influences. Each interview was taped, transcribed, coded, and tabulated.

The third study focused on community-supported agriculture on the Central Coast of California. Community-supported agriculture is an alternative approach to food production and provision in which consumers pay farmers at the beginning of the growing season; in exchange they receive a weekly share of produce. The purpose of this study was to document how community-supported agriculture was being implemented in this area, to assess the extent to which groups practicing community-supported agriculture (CSAS) were meeting the goals ascribed to them in the alternative agrifood movement and to identify the opportunities for and constraints on meeting these goals. In this study we wanted to obtain the perspectives of both producers and supporting community members. For producers, data were collected both through in-depth interviews with twelve community-supported farmers (out of fourteen in the area) and a written questionnaire. Information on member experiences and perspectives was gathered through a written questionnaire included in the members' boxes or sent through the mail. We received 274 responses to the 638 surveys delivered to members, a response rate of 43 percent. In addition, we held three focus groups with seventeen members of five different farms. Focus group members were self-selected by identifying their interest in participating on their written questionnaires.

While the information about alternative agrifood institutions in this book has been gathered from a number of sources using multiple methods, it is less inclusive in its geographic reach. All of the data and examples come from the United States.

Area of Focus: United States and California

This research focuses primarily on alternative agrifood movements in California and in the rest of the United States because of the worldwide economic and political significance of their agrifood systems. The dissemination of the American model of production and consumption to other countries, combined with technological leadership and unchallenged supremacy of the United States in world markets, has "effectively established an international food order under North American hegemony" (Marsden and Little 1990: 26). American leadership in agricultural production volume and sales is beyond dispute. The United States exports far more edible agricultural products than any other country—almost half again as much as the next highest export country, which is France (Food and Agriculture Organization 1996). And it is partly due to this economic power that American food production systems and technologies are promoted and emulated throughout the world.

Within the United States, California possesses the premier food and agricultural system. As the world's sixth-largest economy, with a land mass roughly equivalent to that of the United Kingdom, and home to 34 million people, California is almost more like a country than a state. It has led the nation in agricultural production and income for nearly fifty years, and its agricultural economy ranks sixth among nations as an exporter of agricultural products. In part because of its climate, productive soils, and irrigation system, California ranks first in the nation in agricultural production value for 75 crop and livestock commodities, generating $24.8 billion in sales in 1996 (California Farm Bureau Federation 1998). California agriculture is one of the most diversified in the world, producing over 250 different crop and livestock commodities, with no single crop dominating the state's agricultural economy. Although its 30 million acres of farmland account for only 3 percent of the country's total, it produces 55 percent of the nation's fruits, nuts, and vegetables.

Long held up as an exemplar for the rest of the nation and often the world, California's agrifood system is assuming a leadership role in the domains of sustainable agriculture and community food security as well. Within the state, organic farming is a significant and growing industry, generating $95.1 million in sales in 1995, a 26 percent increase over the previous two years (Torte and Klonsky 1998). California has extensive experience in all aspects of sustainable agriculture. As a result of the organizing efforts of California Certified Organic Growers, as early as 1978 California developed legal standards for organic agriculture in California. This law was used as a model by the group drawing up the rule that became federal policy in 2002. Another institutional marker is that the national office of the CFSC was established and remains in California, and 25 percent of its membership resides there.

California provides an excellent opportunity for studying the possibilities of a movement that combines environmentalism and justice in food and agriculture. Because of the ways in which California agriculture differs from that of America's agricultural "heartland," there may be greater potential in California than in other agricultural regions for the development of alternative movements. Unlike agriculture in many parts of the country, California agriculture has been explicitly capitalist from the start, underscoring many of the contradictions that the sustainability and food security movements address.

From the beginning, California agriculture was based on the intensive extraction of natural resources and the reconfiguring of nature according to the logic of intensive agricultural production for export. California agriculture is based on extensive irrigation systems and the intensive use of fertilizers. The same long growing season and mild winters that enable the high production of so many fruit and vegetable crops also allows pest populations to grow, leading to high rates of pesticide application.

California is also the nation's first and most extensive example of highly concentrated agriculture, with over 50 percent of production controlled by only 10 percent of the farmers by the end of the 1920s (Jelinek 1982). While large-scale agribusiness is a feature of agriculture throughout the country, corporate involvement has tended to be in input, marketing, and processing rather than in direct production. The entry of large corporations in farm production has been the exception in most parts of the United States (Pfeffer 1992), but not in California. Agricultural land ownership has been highly concentrated in the West since the arrival of Europeans, and this concentration led to the creation of a dual system of capitalist farmers and wage laborers (FitzSimmons 1990).

While in most parts of the United States farm production is based on family or tenant labor, California agriculture has always depended on seasonally employed migratory workers (Martin et al. 1988). More than 85 percent of all of the labor that produces the state's crops and livestock is performed by hired workers (Villarejo et al. 2000). California agriculture presents a clear juxtaposition of deep social inequality with unparalleled abundance. Ironically, the farmworkers who produce and harvest California's bountiful crops comprise one of the populations at greatest risk of hunger. Even in the heart of California's most abundant agricultural region, the Central Valley, children go hungry. The low pay, arduous and dangerous working conditions, and lack of employment security have led to persistent farmworker protests over the years, including a successful interethnic coalition that became the United Farm Workers Union (UFW).

Since at least the 1960s activists in California have raised issues about environmental and social problems in their agrifood system. Environmental concerns focused on agrichemical effects on the environment and ground-water depletion. Social concerns included the plight of farmworkers, the distributional effects of irrigation laws, and the poverty and racism that were part and parcel of the agrifood system. This tradition of activism continues to this day. Today, California has a high density of projects and organizations dedicated to sustainable agriculture and community food security. For example, California is the only state that has developed a statewide community food security organization, the California Community Food Security Network. And, of the five regional Sustainable Agriculture Working Groups, only one is based in a single state, the one in California.

Thus, the social and environmental issues of California's industrialized agrifood system, along with a history of social activism may provide a different type of catalyst for change in California than in other American agricultural regions. Alternative agrifood movements may also have a better chance to flourish within the state's complex and diverse demographic and sociopolitical environment. While conventional agricultural interests are powerful in California, they may be less so than in other states where agriculture is a more significant part of the economy. While California is the nation's leading agricultural producer, farming and related activities contribute only about 8 percent of the gross state product and supply about 8 percent of the jobs in the state (Carter and Goldman 1996). Not only is California's political economy relatively less dependent on agricultural production, but California voters tend to be nonrural and liberal. More than 90 percent of the state's population lives in metropolitan areas, and less than one percent of the state's residents are farmers or ranchers. These conditions pave the way for interests beyond those of conventional producers to help shape the agrifood system of the future.

California provides fertile ground for the development of a progressive alternative agrifood movement. The relatively small contribution of agriculture to the state's current economy, a history of diverse agrifood activism, the emphasis on progressive politics and alternative lifestyles, the high level of cultural diversity, and the degree of involvement with sustainability suggest that if an arm of the movement that joins environmental issues and social justice were to develop anywhere, California would be a likely place.

Primary Themes of the Book

To understand these movements we first need to address why they exist. It is clear that the contemporary agrifood system is not meeting people's food security needs at present and because of the progressive damage that conventional practices are doing to the environment, this situation is likely to get worse. The conventional agrifood system therefore needs significant changes in order to achieve ecological soundness and social justice. Conventional agriculture has been largely self-negating, depleting the natural resources upon which agricultural processes depend and thus producing barriers to long-term environmental sustainability and food security. These are the core issues for those involved in alternative agrifood movements. Since they are well documented and articulated in many other places, I only summarize them in Chapter 2, where I also outline the development of concepts and movements centered on sustainable agriculture and community food security.

The drive toward environmental soundness and social equity in the agrifood system must be waged on many fronts. Interactions among the larger environmental, social, and economic systems in which agriculture is situated directly influence agricultural production and distribution. This means that solutions need to be found both on and beyond the farm, and that solutions will be not only technical but also social and political as well. Alternative agrifood movements realize that they need to engage with the agrifood institutions, such as the U.S. Department of Agriculture (USDA) and the land-grant agricultural research system, that have largely configured the current agrifood system. I discuss how the alternative agrifood movement has accomplished this engagement in Chapter 3, in which I review the institutionalization and key features of national and California programs in sustainable agriculture and community food security. Alternative agrifood movements are also developing concrete alternatives to current methods of production and distribution. In Chapter 3, I also highlight some of the alternative production and marketing practices featured by alternative agrifood movements.

What this book contributes to these efforts is an analysis of how alternatives are moving the agrifood system in the direction of environmental soundness and social equity. Through this review it is clear that the movements have made significant progress in developing alternatives to the current agrifood system and in integrating alternative discourses into dominant agrifood institutions. In many instances, they have challenged and are beginning to change the discourse and practice of these institutions. Discursive space has been carved out for sustainability and food security, and research agendas and methods are consequently beginning to change. These incremental changes are setting the stage for even broader and deeper transformations in major agrifood institutions.

Yet although new alternative agrifood discourses are being established, many traditional, conventional agrifood discourses remain. Chapter 4 identifies and examines both these emergent and residual discourses. In some ways, the institutional forms of sustainable agriculture and community food security have been constructed such that their problems are remediable within the structures of existing institutions. These institutions, in turn, shape the accepted frameworks of sustainable agriculture and community food security. Buttel (1997), for example, points out that although the sustainable agriculture movement is based upon broad social values, its effectiveness within traditional institutions is a based upon promoting a set of technical practices institution leaders consider both comprehensible and relatively noncontroversial.

This process occurs without any group necessarily intending it to happen. For example, the focus on natural science and technology can be seen as an accommodation to the institutions in which these approaches have been privileged and with which their scientists and administrators are familiar. Yet developing an environmentally sound and socially equitable agrifood system requires a larger epistemological framework for analysis than that of traditional agricultural science in order to find common ground and see beyond constructed dichotomies such as production and consumption. One approach suggested for analysis and action is a political ecological framework in which causes of and solutions to problems in the agrifood system are seen as both natural and social.

There is a narrow, and permeable, boundary between residual and emergent—the old and the new—discourse and practice, both within agricultural institutions and within the alternative movements themselves. Chapter 5 explores how the movements may be reproducing some of the discursive approaches and ideologies of the dominant agrifood system, such as economic liberalism and individualism, in which nonsustainability and food insecurity are embedded. For example, while farmers may embrace the idea of sustainability, they face the reality of competition; they are driven by the same economic considerations that conventional farmers are. Within the exigencies of the market economy, one must make a profit or get out. Untangling these kinds of Gordian knots requires self-reflection on movement discourses and ideologies.

A crucial discursive step is to clearly define and articulate principles and characteristics of an agrifood system that is based upon environmental soundness and social justice so that the concept of sustainability, for example, cannot be as easily co-opted as it seems to be at the moment. Furthermore, attention needs to be paid to how these principles are interpreted and implemented. For example, many alternative agrifood organizations and programs have vision and goal statements that are broad and inclusive, focusing on environmental soundness and social justice for all food-system participants. In institutions such as sustainable agriculture grant programs or in practices such as alternative marketing strategies, discourse includes everyone. However, these goals tend to narrow as they become operationalized, and at the level of implementation, stakeholder groups such as farmworkers may be excluded entirely.

It is not surprising, then, that sustainability and food security discourse undergoes a narrowing from principles to practices within traditional agrifood institutions. What is possibly more problematic—and also more solvable—is the extent to which this narrowing happens within the alternative movements as well, thereby limiting the claims and changes they attempt to make. This constriction may also be embedded in some of the discourses and ideologies of these movements. Given the central role played by discourse in social movements, it is crucial that this discourse work toward solving rather than reproducing the problems that gave rise to the movement in the first place.

Another issue that bears examination is that of power and participation both in the current agrifood system and in the alternatives promoted by the movement. The primary participants in alternative agrifood movements closely resemble the participants in conventional agriculture in class, gender, and ethnicity. Participants in alternative agrifood movements are caught in power relations and discursive and ideological strangleholds similar to those of conventional agriculture. Chapter 6 addresses issues of authentic democracy as refracted through the prisms of privileged voices, material power, and gender and explores possibilities for deepening and expanding participation in alternative agrifood movements. Given uneven resource allocations among different groups of people, this emergent inclusiveness in turn requires exploring the possibility of democratizing both movements and institutions. So far there has been little discussion of how historically marginalized people can gain access to resources such as education, property, and capital that can give them equal footing in discursive spaces. It is unlikely that a runner who is placed far behind the starting line can catch up with the rest of the field, and this is an issue that needs to be addressed even if it is not clear how it can be resolved.

One of the current major efforts at developing sustainable, just, and democratic agrifood systems focuses on the creation of localized food systems. While these efforts make sense at face value, in Chapter 7 I explore some concerns about the implications of the drive toward food-system localization. These include concerns about the fundamental asymmetries of power within communities and the enormous differences in wealth and resources from one community to another.

In Chapter 8, I look into the current configurations of U.S. food and agricultural policy, including the demographics of the decision-making arrangements that created these policies. After discussing the importance of building broad-based alliances for developing alternative agrifood systems, I address some of the challenges inherent in this kind of effort. I conclude by highlighting emerging alliances for social and environmental justice in the agrifood system.

Now that agricultural sustainability and community food security programs are becoming institutionalized, to what degree should alternative agrifood movements seek further reforms and to what degree should they push for deeper changes in areas such as property relations, participatory democracy, and productive justice? Chapter 9 addresses the very real and troubling tension between reform and transformation faced by all social movements. While building on institutional success, alternative agrifood movements will also need to acknowledge and address the deeper structural and cultural patterns that constrain coordinated efforts to resolve social and environmental problems in the agrifood system. Several steps are crucial to this process: (1) developing a vision for a sustainable and food-secure society; (2) working to understand the causes for a nonsustainable and food insecure society and removing ideological blinders; and (3) realizing that people working together can transform the agrifood system, even at its most fundamental levels. Achieving agricultural sustainability and food security requires both the development of alternative practices and a political struggle over rights, justice, and equity. Whether the future in which we find ourselves is better or worse than the present will depend in large part on the evolving alternative agrifood movements simultaneously prioritizing issues of environmental and human degradation.

This book is an exploration of the concerns, claims, discourses, and practices in the alternative agrifood movement. My intention is to offer information and insights that can contribute to the reflexive efforts of the alternative agrifood movement as it continues to develop. My approach is "critical" in the sense that I attempt to ferret out meanings and connections that may be hidden from view as alternative agrifood advocates pursue the day-to-day actions in which they are engaged. It is *not* critical in the sense of criticizing people or their efforts as they work to change the agrifood system. I have only the utmost respect and esteem for the many people who work against long odds to develop a sustainable food system that provides sustenance for all. Through this work, I hope to provide some illumination along the path toward a more environmentally sound and socially just agrifood system, one that provides for us for both now and indefinitely into the future.

2	perspectives of alternative agrifood movements
	issues and concepts

Throughout human history problems and resulting protests over them have been a feature of agrifood systems, with different issues and social movements rising to prominence at different times. Today people are increasingly aware of the fragile state of America's agrifood system. Contemporary alternative agrifood movements did not, of course, burst forth suddenly, like Athena from the head of Zeus. They have roots in or affinities with previous social movements such as the abolitionist, populist, environmental, antihunger, and food safety movements. While there are a number of contemporary alternative agrifood movements, this book focuses on two of the most comprehensive and prominent, the movements for sustainable agriculture and for community food security. These movements have developed both as legacies of and in reaction to traditional conceptualizations and practices in the American agrifood system. This chapter begins with a review of some of the issues and problems that have inspired contemporary alternative agrifood movements and then highlights some of the agrifood movements of the past. It then turns specifically to the concepts and development of the movements for sustainable agriculture and community food security.

Issues in Sustenance and Sustainability

Today's prominent agrifood concerns are the issues of sustenance and sustainability. To better understand the discourses of sustenance and sustainability, I categorize these issues as centered around three main themes: food, environment, and livelihood and life chances. Food issues include those of food access and hunger, nutrition, and food safety. Environmental issues span a spectrum from depletion of natural resources such as soil and water to the deleterious effects of agrichemicals such as groundwater contamination by fertilizers and pest resistance to pesticides. In the category of livelihood and life chances, I include issues such as the working conditions of farm laborers and the concentrated ownership of farmland and food and agrifood businesses.

Food

No other commodity is more essential than food—like water, it is absolutely required for human survival. Access to water, however, has not yet been determined by one's ability to pay for it, although in some places this is beginning to change. Everyone—regardless of age, gender, ethnicity, or social class—needs to eat in order to live. Yet at least 500 million people, mostly women and children, are chronically undernourished, and many more lack the proper diet for a healthy, active life (U.N. World Food Council 1990). In a world that produces enough food for all, each day forty thousand people die of hunger and hunger-related causes (Speth 1992). While those at greatest risk of hunger are women and children living in rural areas of Asia, Africa, and Latin America, many Americans also go hungry.

Indeed, a defining contradiction of American agriculture has been the persistence of hunger despite its having the world's most productive agrifood system. As mentioned in Chapter 1, American agriculture is legendary in its levels of productivity. The United States produces plenty of food for its own population and enough to support an enormous export program. Yields in most crops have increased dramatically since the first part of the 1900s, and Americans on average spend only about 10 percent of their incomes on food—a much lower percentage than in any other country. We have access to a much more diverse diet than at any point in the past and in many ways are much better nourished than ever before.

Still, many Americans do not have enough to eat. In 1999, 31 million Americans were considered food insecure by the USDA, and food insecurity is on the rise. For example, while in the 1980s there were fewer than thirty emergency food centers in New York, today there are thirteen hundred. Hunger is unevenly distributed among different groups of people. Those most likely to suffer from food insecurity are people of color, the elderly, the disabled, inner-city residents, farmworkers, and children. Of 31 million people considered food insecure, close to 40 percent were children. Sadly, children go hungry even in California's Central Valley, a showcase of modern agricultural productivity. In California—the wealthiest state in the world's wealthiest nation—1.4 million children are hungry or at risk of hunger (True 1992). Since America's ability to produce food is not in question, providing adequate nutrition for everyone clearly involves factors that go far beyond achieving sufficient food production.

Given that food is treated as a commodity, it is axiomatic that the primary cause of food insecurity is poverty. For many, the economic picture in much of both urban and rural America is bleak, with wages often too low to keep many workers out of poverty, particularly women and ethnic minorities. In rural America in the late 1980s, for example, one-fourth of children lived in poverty, even though 75 percent of them lived in a household with at least one working adult (O'Hare 1988). Since the need for food is related to biology, not economics, a person with a low income needs to spend a higher percentage of his or her income to meet basic food needs than does a middle- or high-income person. In addition, poor people often pay higher prices for their food. Because of supermarket redlining in low-income communities, in many poor neighborhoods the only food stores are small businesses whose low volume of sales means that they cannot acquire food at low cost and therefore cannot charge low prices to their customers. In addition, even in supermarkets, food prices in low-income communities are often higher than those in other areas.

While hunger is the key problem for many Americans, for others it can be the overabundance of food. In the United States rich and poor alike struggle to escape the new plague of diseases caused by consuming too much and the wrong kinds of food. Many of the major American diseases are related to dietary excesses and imbalances. For the two-thirds of Americans who neither smoke nor drink excessively, "One personal choice seems to influence long-term health prospects more than any other—what we eat" (U.S. Department of Health and Human Services 1988). For example, obesity and physical inactivity account for more than 300,000 premature deaths in the United States each year, second only to deaths related to tobacco use (National Center for Chronic Disease Prevention and Health Promotion 2000). Childhood obesity has become a national epidemic. Dietary excess has long been associated with the leading causes of death in the United States such as cardiovascular diseases (coronary artery disease, stroke, and high blood pressure), cancers (colon, breast, and prostate), and type 2 diabetes (McGinnis and Foege 1993). The food industry spends huge amounts of money each year to get Americans to buy foods not based on their nutritional or health value but on their value as contributors to food industry profit margins.

As a result, Americans are confronted with a bewildering array of food choices, each more processed and chemicalized than the one before. Both farm-fresh and processed foods may be contaminated with pesticide residues. Processed foods are often stripped of nutritional content and supplemented by chemical additives and are often high in fat and sodium. On the other hand, foods with less processing may be more likely to contain microbial pathogens such as salmonella (Leon and Smith DeWaal 2002). One in four Americans suffer from some form of food poisoning each year, and five thousand die as a result of eating contaminated food. We have all read about the tragic cases of children dying from drinking fresh fruit juice or eating undercooked hamburger. In the meat industry, one cause of the proliferation of pathogens is the subtherapeutic use of antibiotics to increase rates of growth in livestock. Just as pests become resistant to pesticides, bacteria become resistant to antibiotics. This reduces the efficacy of antibiotics for fighting disease in both livestock and humans.

Food in the United States has become almost a negation of itself, either because it is absent, harmful to our health, or because it is virtual in the sense that any nutritional content has been neutralized. Often crops are no longer recognizable final consumer goods but serve as raw materials for the food industry, where it has become commonplace to substitute technically developed products for tropical crops in food manufacture. For example, high-fructose corn syrup has replaced sugar in many items such as soft drinks, cookies, and gum. By 1985 the use of corn sweeteners surpassed that of cane and beet sugar (USDA 1996). These changes represent a confluence of the fiction and absence of food. Lower demand and prices for products previously imported from impoverished countries contributes to further impoverishment and food insecurity abroad and poorer health at home. These products are also extremely expensive relative to raw agricultural products, thus contributing to problems of food insecurity in richer countries as well. Problems such as hunger, diet-related disease and mortality, and food safety are clearly urgent and immediate. The agrifood system is also replete with environmental problems that seem less immediate, but are no less crucial.

Environment

No other commodity is as "natural" as food. While all commodities begin and end in "nature," this is particularly clear in the case of food and agriculture. Since agriculture depends upon the primary appropriation of nature, it is a special case of the intersection of production and environment (Mann and Dickinson 1980). Even in its industrialized form, agriculture remains dependent upon natural resources and processes such as soil, water, and weather. Rates of production are limited by natural constraints such as growth cycles, weather, and length of day. Agricultural production begins in nature as resources are transformed into food. It ends in nature as waste products and pollution from materials applied to it in attempts to control the constraints of nature (e.g., pesticides and fertilizers). The production and distribution of food is the outgrowth of a highly visible, intensive relation between people and the environment.

Agriculture's direct dependence upon natural resources and processes makes it impossible to obscure environmental destruction in the agrifood system. In places where agriculture has produced abundance, it has often done so at the cost of environmental quality. Much of this destruction has been concealed in technological innovations such as new developments in fertilizers, pesticides, and cultivation techniques that have enabled continued increases in production. Yet these innovations present their own problems—pesticides produce pests, irrigation produces groundwater depletion, cultivation produces soil erosion.

The discovery of insecticides based on synthetic organic compounds around the time of World War II greatly increased the use and consequences of pesticides in agriculture. In a very short time they were being used on almost every crop in most countries of the world (Conway and Pretty 1991). Any increased application of pesticides intensifies future needs for more chemical toxins, as pests develop resistance to standard preparations. Losses to pest resistance have already severely reduced or destroyed agricultural industries in several parts of the world, since pest resistance renders pesticide application a self-negating process. In California, for example, pesticides are responsible for the growth of secondary pest populations, which now comprise twenty-four of the state's twenty-five major crop pests (Metcalf and Luckmann 1982). While pesticide use in the United States increased 1,000 percent between the 1940s and the 1980s, crop losses to insect pests also increased by almost 50 percent (Pimentel et al. 1991).

The pesticides used extensively in modern agricultural production damage wildlife, beneficial insects, ecosystems, and humans. Since less than 0.1 percent of pesticides applied in the United States actually reach the pests to which they are targeted (Pimentel and Levitan 1986), pesticides end up in the bodies of wildlife or the water people drink. Agriculture is the most prominent cause of species endangerment in this country (U.S. Forest

Service 1994). Similarly, non-point-source pollution from agriculture is the major contributor to water-quality problems in America's surface water, and agriculture contributes to pollution in over one-half of the assessed streams, rivers, lakes, and reservoirs suffering impairments (House 1995). The herbicide atrazine, a carcinogen and endocrine disrupter, causes more health violations in tap water than any other chemical regulated by the Environmental Protection Agency (Environmental Working Group 2000). Pesticide contamination can remain long after the compound is no longer used. In California, for example, the long-banned pesticide DBCP, one of the most potent carcinogens known, still contaminates the water of 1 million Californians at levels that are almost three hundred times the "safe" level for infants and children (Environmental Working Group California 1999). According to the Office of Technology Assessment (1995a: 8–9), "Overall, water quality suffers most from its association with agriculture. Agriculture ranks as the primary contributor to today's surface water quality problems, principally through sediment deposition and agrichemical runoff from dryland and irrigated systems." Soil compaction caused by heavy cultivation, land salinization caused by salt build-up from irrigation, and changes in soil biology caused by fertilizer use also threaten agricultural productivity.

In addition to resource degradation, resource depletion is a major problem. Approximately one-third of the original topsoil has been removed from U.S. cropland in the past two hundred years (Pimentel et al. 1994), and much of U.S. cropland erodes at rates that exceed government-established tolerance levels. The extensive use of groundwater for irrigation has meant that declining water tables have become common in many agricultural regions. As early as the 1970s, agriculture was depleting groundwater at the rate of 21 billion gallons per day (U.S. Water Resources Council 1978). Resource depletion and degradation have caused the abandonment or threaten to cause the abandonment of farming systems through groundwater depletion, soil salinization, and unmanageable pest problems caused by pesticide use (Lockeretz 1989). In the United States, an estimated one billion hectares of arable land has been lost to erosion, salinization, and waterlogging (Pimentel et al. 1976). Worldwide, these same processes are causing an irretrievable loss of an estimated 6 million hectares per year (Pimentel 1993).

Even this limited number of examples illustrates the severity of environmental problems in the agrifood system. Not only is agriculture responsible for pollution, but agricultural practices are contributing to the destruction of the environmental conditions of production upon which agriculture itself depends.

Livelihood and Life Chances

In some ways, the current agrifood system also destroys the dignity and opportunities of particular groups of people. Inequitable social relations are deeply embedded in food and agriculture systems throughout the world. Gender and racial oppression have functioned as primary organizing principles, and labor exploitation is the rule.

The American agrifood system is one that embodies and has depended upon extremely unequal material and social relations among groups of people. For example, in California some of the richest agricultural areas are home to some of the poorest people in the entire United States. In fact, increases in income from agriculture have been associated with increasing levels of poverty (MacCannell 1988). Farmworkers have the lowest family income of any occupation surveyed by the U.S. Bureau of the Census. Half of U.S. farmworker families have incomes below the poverty level, with the median family income between $7,500 and $10,000 a year (GAO 1992a). This figure is particularly striking in that 1.5 percent of U.S. farms with the highest sales employ over half of the farm labor (Slesinger and Pfeffer 1992). And at the end of the workday, many farmworkers do not have a home to which to retreat. The only national data on farmworker housing show that in 1980, housing was available for only about one-third of the estimated 1.2 million migrant farmworkers who needed it (GAO 1992a). Most farmworkers live in extremely overcrowded conditions; others end up sleeping in caves, under bridges, or in cardboard shanties. Many still work without access to restrooms or fresh drinking water, although access to these so-called amenities was a central goal of labor-organizing efforts as far back as the early 1900s.

Difficult working conditions are endemic throughout the food and agriculture sector, not just in the fields. Workers in the produce and meat-processing industries are often poorly paid, seasonally terminated, receive no benefits, and work under miserable conditions. In the 1980s, Iowa meat-packing industry wages decreased regularly, and 49 percent of Iowa meat-packing workers suffered work-related injuries or illnesses in 1989 (Senate 1990). These plants are increasingly staffed by recent immigrants who have few income-earning options and little ability to protest their working conditions. In Hamlet, North Carolina, twenty-five workers were killed and forty-nine were injured when they could not escape a fire the Imperial chicken-processing plant because the emergency exit doors were locked. Congressman George Miller summed up the situation by saying that this was an industry that decided to subsidize its profits "with the broken lives, limbs, lacerations, and decapitations of their workers" (House 1991: 16). These kinds of working conditions are enabled by the highly uneven distribution of control and ownership in the U.S. food and agriculture industry, a level of concentration that affords workers little power to change either their working conditions or their jobs.

The U.S. food and agriculture system is highly concentrated in production, retailing, and land ownership. At the level of production, only 7 percent of American farms received 60 percent of the net cash farm income in 1992 (USDA 1994). As for marketing, at the beginning of the 1990s two companies controlled 50 percent of grain exports; three companies slaughtered nearly 80 percent of the beef; four companies controlled nearly 85 percent of the cold cereal market; and four companies milled nearly 60 percent of the flour (Krebs 1991). Similarly, the food service industry is dominated by only three companies. There is also a long-term trend toward larger and fewer grocery stores across the United States. Supermarket chains dominate grocery retailing, accounting for four out of every five dollars spent in retail food stores (Geithman and Marion 1993). In 2000, the top five food retailers (Kroger, Albertson's, Wal-Mart, Safeway, and Arnold) controlled 42 percent of the market (Hendrickson et al. 2001).

Land ownership is also highly concentrated. Only 5 percent of American landowners own 80 percent of the land (Hansen 1999). Compare this to the situation in Brazil—a country considered to be an extreme case of land concentration—where 3 percent of the landowners own 56 percent of the arable land. And although African Americans, Latinos, and Asian Americans have been essential to the productivity of American agriculture, they are much less likely than whites to be farm operators and much more likely to be farmworkers. Although they comprise nearly 25 percent of the population, nonwhites operate a mere 2 percent of the farms in the United States (Census Bureau 1987). Even in California, an ethnically diverse state where 43 percent of the population is nonwhite, less than 7 percent of farm operators are nonwhite (Census Bureau 1987). In contrast, California's farm labor force is composed almost exclusively of ethnic minorities (Peck 1989).

Alternative agrifood movements have arisen in response to these kinds of food security, environmental, and livelihood problems in the American agrifood system. While agrifood system problems may be more severe and more publicized today, neither the problems nor the organizing around agrifood issues is new in America.

Alternative Agrifood Movements in Historical Context

For over a century, conditions in American food and agriculture have led to or been associated with resistance movements such as the populist, environmental, antihunger, and food safety movements.

Issues related to family-farm viability and market concentration were raised by the agrarian populist movement of the late 1890s.[1] During the thirty-year period following the end of the Civil War in 1865, agricultural production increased dramatically. Agricultural export earnings went from $79 million in 1865 to $242 million in 1881 for crude foodstuffs (Havens 1986). Post–Civil War industrialization and government monetary policy eventually produced a situation in which farmers experienced generally declining prices while the costs of shipping their products, purchasing farm inputs, and obtaining necessary credit increased (Adamson and Borgos 1984). This led to the rise of political action intent upon easing the plight of farmers and to the creation of political parties such as the Populists and the Greenbacks. These parties saw the power of the banks, railroads, and monopolies as central to the economic problems experienced by farmers. Their platform included regulation of the railroads, expansion of the national money supply (to lower interest rates), legal recognition of trade unions, and taxation on speculative real estate profits. Agrarian populism was revived in the late 1960s in defense of the family farm and traditional rural communities (de Janvry 1980). The neopopulists denounced the technological, public-policy, and market advantages that large-scale agriculture enjoyed over small-scale farming. In the late 1970s agrarian activism coalesced into the American Agriculture Movement, which promotes the family farm and the importance of agriculture to U.S. economic security (Browne 1988). In 1978 they organized some of the largest farm demonstrations in history when thirty thousand farmers marched in Washington, D.C., to protest American farm policies.

Environmental degradation in agriculture also met with early social criticism, which addressed resource problems in "modern" agriculture at least since the closings of the commons in the 1700s and through the early

---

1. See Stock 1996 and McConnell 1959 for two quite different histories of the American populist movement.

British and U.S. conservationist movements of the 1800s. As agricultural productivity began to decline dramatically in the early nineteenth century both in Europe and the United States, technological efforts to overcome the constraints of nature included chemical and mechanical means, such as the development of artificial fertilizers and tillage equipment. These solutions, however, led to further natural resource problems and were widely recognized and criticized (Foster 1997). For instance, in his *Lectures on Modern Agriculture* of 1859, the eminent soil chemist Justus von Liebig considered the agricultural systems of the time to be forms of "robbery" in which the "conditions of the reproduction" of the soil were destroyed. American economist Henry Carey wrote in 1858 that "Man is but a tenant of the soil, and he is guilty of a crime when he reduces its value for other tenants who are to come after him." During this same period, Karl Marx was also highly critical of the soil-destroying dynamic of capitalist agriculture and believed that humans must cease the wanton destruction of nature. Marx considered "the soil and the worker" to be the fundamental sources of wealth (Marx 1976). Karl Kautsky (1988) understood the concept of diminishing returns to increased agricultural inputs, writing in *The Agrarian Question* of 1899 that artificial fertilizers could only temporarily enrich the soil, not prevent its eventual impoverishment. These cogent early analyses about the sources and dynamics of agricultural resource problems prefigure contemporary concerns about agricultural sustainability. It was not until the 1962 publication of Rachel Carson's *Silent Spring*, which raised previously unasked questions about harmful effects of pesticides, that agriculture was featured in the contemporary environmental movement.

Health and food safety movements also have a long history in the United States. As early as the 1830s, for example, vegetarians protested public health recommendations for a heavily meat-based diet (Belasco 1989). At the end of the nineteenth century, the industrialization of the food system gave rise to efforts at reforming such practices as food adulteration (Guthman 1998). For example, dairies artificially colored milk because it turned blue as a result of cows being fed with byproducts of distilleries, and bakeries were accused of adding nonfood substances to their bread to cover up impurities and make it heavier and whiter (Leon and Smith DeWaal 2002). Upton Sinclair's graphic account of the meat-packing industry in his novel *The Jungle* caused an outcry that led to regulations aimed at improving food safety and controlling fraud in the early 1900s. Several decades later, food issues came to the fore again during the time of the civil rights, free speech, and antiwar movements. In the 1960s, food activism was broad-based, including fasts against the war, interracial dining at segregated restaurants, and consumer boycotts in support of agricultural workers (Belasco 1989). This was also the point at which the movement for organic food began to flourish as part of a growing rejection of "mainstream foodways" (Belasco 1989).

During this same period of social activism and growing out of the civil rights movement, the farmworkers' movement focused on working conditions, wages, and pesticide exposure. The earliest agricultural labor movements, of course, were the antislavery movements. These were followed by movements focusing on migrant workers during the Depression. However, reforms in farm labor conditions won by these movements were stalled in 1942 with the advent of the federal *bracero* program (Mooney and Majka 1995). This program brought temporary Mexican workers to the United States to work in the fields. This process ensured an oversupply of workers, which in turn essentially eliminated the ability of workers to organize. The end of the *bracero* program in 1964 coincided with the surge of the civil rights movement. In 1965 Cesar Chavez, coming out of his background as an organizer for a community action agency, became the first leader of the UFW (Mooney and Majka 1995). During the 1970s consumers supported the union through a boycott of table grapes, head lettuce, and Gallo brand wines designed to apply pressure for legislation that would give farmworkers the right to organize without threat of retaliation. This led in 1975 to the passage of the Agricultural Labor Relations Act (ALRA) in California, which protects the rights of farm laborers to organize and choose their own representatives and created the Agricultural Labor Relations Board (ALRB). During the 1980s, however, Board appointments made by Governor Deukmejian, who sided with growers, rendered the ALRB virtually useless in mediating disputes between growers and workers. This, combined with internal conflict within the UFW, led to a precipitous decline in union membership. In the early 1990s the UFW, under new leadership, undertook a new campaign to unionize strawberry workers in California. This effort, which involved forging an alliance with consumers and the AFL-CIO, focused on both labor and environmental issues. While the history of most farm labor organizing involves opposition between growers and workers, in other areas such as the Midwest, where the labor structure of agriculture is different, there have been efforts in which workers and small farmers have joined forces to advocate changes in food-processing industries. For example, the Farm Labor Organizing Committee, recognizing that working conditions were set more by processors than by individual farmers under contract to the processors, worked together with farmers, consumers, other unions, and churches to eventually win better contracts for the workers (Mooney and Majka 1995).

The alternative agrifood movements of today carry on these traditions of activism and have made their voices heard. In the late 1970s debates over farm policy were no longer carried out by exclusively agricultural interests, but now included groups representing the "New Agenda" (Paarlberg 1980), such as those concerned about the environment, food safety, nutrition, and government expenditures. An increased public concern with environmental degradation, for example, led to environmental issues becoming part of farm bill discussions for the first time.[2] Many of the groups that became active during that time are even stronger today.

Some focus on single issues such as food safety, farmland protection, pesticides, farm labor, or organic farming. Others, like the Pesticide Action Network North America, focus on "crossover" issues such as pesticides and farmworker health. Here I focus on two prominent, broad-based alternative agrifood movements, the movement for sustainable agriculture and the movement for community food security. While there is a confluence of issues addressed and strategies used by these movements, each one embodies a somewhat distinct conceptual and political history. Traditionally, advocates of sustainable agriculture have focused more toward on rural and production issues, while community food security proponents have concentrated more on urban and consumption issues. For example, in a review of the literature on agricultural sustainability, Lockeretz (1988) identifies the problems addressed as environmental contamination by pesticides, plant nutrients, and sediments; loss of soil and degradation of soil quality; vulnerability to shortages of nonrenewable resources, such as fossil energy; and low farm income resulting from depressed commodity prices in the face of high production costs. Problems addressed by the community food security movement, on the other hand, tend to focus more on food and nutrition issues. These include hunger and poor nutrition, high rates of diet-related disease, unprecedented demand on the charitable food sector, abandonment of inner cities by the supermarket industry, the decline of local food systems, and the absence of community or individual empowerment (Fisher and Gottlieb 1995).

The next sections review the development of the concepts of sustainable agriculture and community food security. They illustrate some of the ways these movements differ from conventional agriculture and antihunger paradigms and narrate the rise of these important movements, both nationally and in California.

2. The farm bill is a piece of omnibus legislation first developed during the New Deal. Renewed every five to six years, it covers subsidies to agriculture, research priorities, and funding for food and agriculture and for food stamps.

Sustainable Agriculture

The movement for sustainable agriculture combines issues central to the concerns of populists and environmentalists. In this section I briefly describe the history of the concept of sustainability, discuss the evolution of agricultural sustainability, and look at the creation of organizations dedicated to the promotion of sustainable agriculture.

Concepts of Sustainability

Sustainability binds together otherwise disparate thinking and concerns about the environment and the economy. The 1980s brought about a reorientation of environmentalist thinking in which sustainability became the key concept in development planning and economics (Turner 1988). Dicks (1992) observes that sustainability emerged as a theme that unified environmental concerns voiced during the debates on environmental legislation throughout the 1980s. American interest in sustainability issues since World War II falls into three distinct periods (Ruttan 1992). During the first of these, in the late 1940s and early 1950s, Americans questioned the adequacy of natural resources to sustain growth. During the second period, in the late 1960s and early 1970s, concerns mounted about the external costs (i.e., costs that are not included in private business calculations) of commodity production, such as industrial pollution, and pesticides in food. The third and present wave of sustainability concerns began in the mid 1980s and focused on transnational issues such as global warming, biodiversity, ozone depletion, and acid rain.

Because of these concerns, environmentalists often found themselves at odds with the goals of economic developers, even when those goals included humanistic objectives such as increasing food security. Most development plans for alleviating poverty and hunger were based on models that involved even greater depletion or degradation of natural resources. During this time the concept of sustainable development emerged in an attempt to resolve the perceived contradiction between environmental conservation and economic growth. Sustainable development has been defined as "a strategy for improving the quality of life while preserving the environmental potential for the future, of living off interest rather than consuming natural capital. . . . The key element of sustainable development is the recognition that economic and environmental goals are inextricably linked" (National Commission on the Environment 1993: 2). The publication of *Our Common Future* by the World Commission on Environment and

Development (1987) catalyzed a new wave of thinking and action around merging priorities of the North and the South under the rubric of sustainability. Less and less were environmentalists inclined to villainize those in the South as destroyers of habitats; instead they searched for ways to blend environmental conservation with economic development. The International Union for the Conservation of Nature, for example, promoted "eco-development," regional economic development based on sustainable use of physical, biological, and cultural resources, as an attempt to fuse conservation and development. The World Commission on Environment and Development (1987: 9) points out that "sustainable development is not a fixed state of harmony, but rather a process of change in which the exploitation of resources, the direction of investments, the orientation of technological development, and institutional change are made consistent with future as well as present needs." This has led to criticism that sustainable development is more of a strategy for "sustaining development" rather than supporting natural and human life.

During this period sustainability came to be accepted as a mediating term that bridged the gap between developers and environmentalists (O'Riordan 1988). Traditionally, in their quest for economic improvement, developers paid little attention to the environmental consequences of their projects. As a crucial element of economic development and stability, sustainable agriculture is a derivative and subset of sustainable development. The United Nations' Agenda 21, adopted at the Earth Summit in 1992, promoted sustainable agriculture and rural development as a plan for meeting food needs without further degrading natural resources. The 2002 World Summit on Sustainable Development continued on the path of raising awareness of connections between poverty and resource degradation.

The Development of Sustainable Agriculture in the United States

In America, the New Deal farm policies of the 1930s made some overtures toward the importance of soil conservation, although they were developed primarily as supply management programs.[3] Despite early recognition of problems of soil erosion and pesticide contamination, American agricul-

---

3. For example, the purpose of the Soil Conservation and Domestic Allotment Act of 1936 was to link production control to soil conservation objectives (Robinson 1989). Through this program, benefits were paid to farmers for planting soil-building crops instead of surplus crops. The Act also created the Soil Conservation Service to give technical aid in developing soil conservation plans and practices.

ture increasingly adopted input-intensive production regimes, driven in part by government subsidies. Then, during the energy crisis of the 1970s, the price of petroleum-based farm inputs (fuel, pesticides, and fertilizers) rose with the price of oil. Subsequently, as people began to question the energy intensification of industrialized agriculture and reconsider the deleterious effects of increasing pesticide use, the contemporary concept of agricultural sustainability first emerged (Buttel et al. 1990).

Interest in and activities around sustainable agriculture grew in the early 1980s, fueled by concerns about resource depletion (such as groundwater overdrafts and soil erosion and salinization), environmental contamination (such as nitrates in groundwater), water quality damage through sedimentation, direct and indirect pesticide poisoning, wildlife habitat loss, and the diminishing increase in marginal yields in response to additional inputs such as fertilizers. These issues contributed to a growing sense that modern agricultural production could not be sustained indefinitely. A number of influential writings on the need for a more sustainable agricultural system were published. For instance, Wes Jackson's *New Roots for Agriculture* introduced many people to the idea that agricultural production could be modified to work with rather than against the environment (Jackson 1980). And despite increasing criticism of the USDA's contributions to agricultural problems, ironically it was a USDA report that provided a spark to the incipient sustainable agriculture movement. This 1980 publication, *Report and Recommendations on Organic Farming*, provided evidence of the existence and efficacy of organic farming enterprises in the United States.

Following in the pioneering organic farming and gardening tradition of his father, J. I. Rodale, Robert Rodale expanded the concept and brought it to a larger audience with articles such as "Breaking New Ground: The Search for a Sustainable Agriculture" (Rodale 1983). Various concepts such as organic farming (based on specific agricultural practices), biodynamic agriculture (based on philosophy), and agroecology (based on environmental science) emerged in the ensuing discussions (Dahlberg 1991). This agricultural movement is referred to by several different names, depending upon the period in which people wrote about it and the aspects they chose to emphasize. Prominent among these names are "low-input agriculture," "ecological agriculture," and "organic farming." The term "sustainable agriculture" has emerged as the most prevalent, in part because it has been accepted by national and international agricultural agencies.

Another aspect of the agricultural sustainability movement was economic. The farm financial crises of the late 1980s, precipitated by an overexpansion of production, increased interest among farmers and government agencies in finding alternatives to more conventional agricultural practices. To meet expanding global demand during the 1970s, American farmers increased production financed largely by credit. Then 1981 brought record high interest rates and low commodity prices caused by bumper crops. Although U.S. farm production was at its highest level in history, 1982 was the worst year for farm income since 1932. The 1984 global recession led to a further decrease in demand and therefore prices, leading to record expenditures on U.S. government farm programs. In 1983 USDA expenditures amounted to 413 percent of net farm income. At this time American farmers' largest production expense was interest payment on farm loans (Wilkening and Gilbert 1987). When demand fell, farmers saw not only their markets dry up, but the value of their land decline. As a result they could not make their loan payments. Farm bankruptcies were at their highest level since the Depression.[4] Publicity about farmers losing farms that had been in their family for generations, along with stories of farmer suicides, tugged at the heartstrings of both rural and urban Americans. With his publication of *The Unsettling of America* in 1988, Wendell Berry persuasively argued that agriculture was as much about human culture and values as it was about producing food (Berry 1988). This book, along with the farm crisis, led to the general public's embracing the ideal of the small family farm. A government report framed sustainable agriculture as the fourth major era in agriculture (following the horsepower, mechanical, and chemical eras), stating that the effects of this new era could be more profound than those of previous agricultural revolutions (GAO 1992b).

Comparing Sustainable Agriculture and Conventional Agricultural Approaches

Several studies have examined how sustainable (or alternative) agriculture differs from the conventional agrifood paradigm. Beus and Dunlap (1990), for example, compared the writings of six influential proponents of alternative agriculture with six leading proponents of conventional agriculture in order to clarify core beliefs and values embodied within each perspective (Table 1). Through this review, six major distinctions between alternative and conventional agriculture emerged: centralization vs.

---

4. Pfeffer (1992) points out the similarities of the farm financial crisis of the 1870s and the 1980s. Both had to do with expansion of production financed by loans and the subsequent collapse of the market for agricultural products.

decentralization, dependence vs. independence, competition vs. community, domination of nature vs. harmony with nature, specialization vs. diversity, and exploitation vs. restraint. With regard to the first three distinctions, the perspectives of alternative agriculture closely parallel those of earlier agrarian movements, which traditionally resisted unbridled growth in agriculture, advocated decentralized production and marketing, and affirmed farming as a meaningful and virtuous way of life. Today's alternative agriculturalists, however, are much more concerned with the environmental aspects of agriculture than their agrarian populist predecessors were.

This interest in the environment is a primary difference between alternative and conventional agriculture. Alternative agriculture emphasizes cooperation with nature while conventional agriculture has treated nature as something to overcome. Similarly, while alternative agriculturalists see on-farm diversity (of crops and livestock) as a linchpin of sustainable practice, conventional agriculturalists see specialization and monoculture as essential to efficiency and productivity. On the subject of resources, alternative agriculturalists condemn current practices (e.g., soil erosion and groundwater depletion) as borrowing from the future, while conventional agriculturalists believe that only by harnessing resources with advanced technology will we be able to feed the world and enjoy the affluence achieved through increased agricultural production.

Dahlberg (1991) also contrasts conventional and alternative agriculture, claiming that the debate represents both different perspectives on the future direction of agriculture and a clash between different worldviews (Table 2). In his discussion, a tension exists between institutions, such as government agencies and international bodies, and the advocates of alternative agriculture. For conventional agriculture, the measure of success is high productivity and profits. Alternative agricultural groups, in contrast, privilege small-scale production units, healthy communities, and social equity. Another key difference between conventional and alternative agriculture is one of different planning horizons. While alternative agriculturalists take the long view, conventional agriculturalists often use short time frames that correspond to federal policymaking cycles of only a few years duration. In general, conventional agriculture's approach to agricultural policy typically reflects satisfaction with or acquiescence to current policy, such as commodity subsidies and tax laws (although some conventional agriculturalists advocate a more neoliberal trade regime without subsidies). Alternative agriculture, on the other hand, is by definition interested in reforming the agricultural policies that contribute to nonsustainable practices. For Dahlberg, while conventional agriculturalists are not averse to using less

Table 1 Key elements of competing agricultural paradigms according to Beus and Dunlap

Conventional agriculture	Alternative agriculture
Centralization	**Decentralization**
National/international production, processing, and maketing	More local/regional production, processing, and marketing
Concentrated population; fewer farmers	Dispersed populations; more farmers
Concentrated control of land, resources, and capital	Dispersed control of land, resources, and capital
Dependence	**Independence**
Large, capital-intensive production units and technology	Small, low-capital production units and technology
Heavy reliance on external sources of energy, inputs, and credit	Reduced reliance on external sources of energy, inputs, and credit
Consumerism and dependence on the market	More personal and community self-sufficiency
Primary emphasis on science, specialists, and experts	Primary emphasis on personal knowledge, skills, and local wisdom
Competition	**Community**
Lack of cooperation; self-interest	Increased cooperation
Farm traditions and rural culture out-dated	Preservation of farm traditions and rural culture
Small rural communities not necessary to agriculture	Small rural communities essential to agriculture
Farm work a drudgery; labor an input to be minimized	Farm work rewarding; labor an essential to be made
Farming is a business only	Farming is a way of life as well as a business
Primary emphasis on speed, quantity, and profit	Primary emphasis on permanence, quality, and beauty
Domination of nature	**Harmony with nature**
Humans are separate from and superior to nature	Humans are part of and subject to nature
Nature consists primarily of resources to be used	Nature is valued primary for its own sake
Life cycle incomplete; decay (recycling wastes) neglected	Life cycle complete; growth and decay balanced
Human-made systems imposed on nature	Natural ecosystems are imitated
Production maintained by agricultural chemicals	Production maintained by development of healthy soil
Highly processed, nutrient-fortified food	Minimally processed, naturally nutritious food
Specialization	**Diversity**
Narrow genetic base	Broad genetic base
Most plants grown in monocultures	More plants grown in polycultures
Single-cropping in succession	Multiple crops in complementary rotations
Separation of crops and livestock	Integration of crops and livestock
Standardized production systems	Locally adapted production systems
Highly specialized, reductionistic science and technology	Interdisciplinary, systems-oriented science and technology
Exploitation	**Restraint**
External costs often ignored	All external costs must be considered
Short-term benefits outweigh long-term consequences	Short-term and long-term outcomes equally important
Based on heavy use of nonrenewable resources	Based on renewable resources; nonrenewable resources
Great confidence in science and technology	Limited confidence in science and technology
High consumption to maintain economic growth	Consumption restrained to benefit future generations
Financial success; busy lifestyle; materialism	Self-discovery; simpler lifestyle; nonmaterialism

Source: Beus and Dunlap 1990:598–99.

environmentally damaging inputs, they subscribe to the idea that people must dominate nature in order to achieve prosperity. Alternative agriculturalists seek a mutually beneficial relationship with nature, believing that human and environmental well-being are interdependent.

While the proponents of both conventional and alternative agriculture embrace some aspects of science and technology, their fundamental approaches are different. What distinguishes the two approaches within a scientific framework is that conventional agriculture tends to be reductionist and single-discipline oriented, while alternative agriculture emphasizes interdisciplinary, whole-farm systems, and often localized research approaches. Since science and technological approaches have been responsible for the introduction of some nonsustainable practices, however, another segment of the alternative agriculture movement is much more circumspect about the role science and technology can play in developing sustainability. Some in this group believe that scientific research must include broader social and ethical criteria; they typically call for a basic restructuring of the agricultural research and extension systems. Others go even further, questioning the relevance of science and technology and citing the inability of the agricultural sciences to resolve fundamental problems in the social and legal systems that led to nonsustainable agriculture in the first place. Dahlberg points out other differences within the alternative agriculture movement itself. For example, one wing of the alternative agriculture movement seeks more basic political changes, such as restructuring land tenure arrangements, while others are content to work toward changes in farming practices. These differences are backgrounded, however, by the extent of agreement on issues within the sustainable agriculture movement in the United States.

The Organizations of the Sustainable Agriculture Movement

Farmers, consumers, development planners, university researchers and educators, policymakers, and environmentalists are all engaged in the sustainable agriculture movement. Organizations active in sustainable agriculture include grower groups such as the California Certified Organic Farmers, the New England Organic Farmers Association, and the Biodynamic Agriculture Association. Increasingly, traditional farmer organizations such as the National Farmers Union show interest in the methods and policies of sustainable agriculture. Established environmental groups such as the Sierra Club, the Natural Resources Defense Council, and the Environmental Defense Fund have been active in the analysis and development of sustainable agriculture. Private, nonprofit organizations such as the Land

Table 2 Comparison of conventional and alternative agriculture

Conventional agriculture	Alternative agriculture
Time Frame Two- to five-year planning horizon used by policy makers	Future decades and centuries
Policy Approach (none listed)	Policy changes focused on reorganization within institutions, e.g., reform of agricultural subsidies and tax laws (Policy changes focused on basic reform, e.g., land tenure and terms of trade)
Approach to Science and Technology Subject of investigation is agronomic factors, excluding social or structural issues Methods are reductionist and based on single disciplines	Faith that current scientific and technological approaches can produce sustainable outcomes (Science and technology can have a positive role only if significantly restructured within a broadened ethical and social framework) Interdisciplinary, systems-based, and localized approaches (Need to examine fundamental discontinuities between social and legal systems and natural systems) Requires fundamental restructuring of current specialization of top-down research, education, and extension services
Goals of Sustainability Maintaining productivity while using environmentally damaging inputs	Better integration of individuals, communities, and nature through socially just and regenerative system
Measures of Success Narrow economic of productivity criteria	System health criteria include economics, ecology, ethics, and equity Health of agriculture depends on diverse, healthy rural landscapes and communities, which in turn depend on healthy agriculture
Visions of the Future Growth and prosperity of urban society depends on application of science and technology to increase human domination of nature	Recognition of humankind's dependence on natural systems
	Need for smaller-scale social and technological systems built around healthy local communities and agroecosystems

Source: Derived from Dahlberg 1991.

Institute in Kansas, the Center for Rural Affairs in Nebraska, the Institute for Alternative Agriculture in Maryland, the Committee for Sustainable Agriculture in California, the Rodale Research Institute in Pennsylvania, and the Center for Science in the Public Interest in Washington, D.C., devote programs to pursuing a sustainable agriculture. Citizen groups such as Mothers and Others Against Pesticides and the Humane Society focus on specific issues within the overall sustainability theme. In California there is even a group of public and private grantmakers, the Funders Agriculture Working Group, whose mission is to promote a sustainable agriculture and food system in the state. While the 1996 directory of American organizations in the sustainable agriculture movement lists profiles of over seven hundred groups (Sustainable Agriculture Network 1996), there is one umbrella organization in the United States.

The National Campaign for Sustainable Agriculture is a nonprofit organization created in 1994 to coordinate unified action within the sustainable agriculture movement. The organization is "dedicated to educating the public on the importance of a sustainable food and agriculture system that is economically viable, environmentally sound, socially just, and humane." Focused primarily on federal policy, the Campaign works with regional organizations to analyze policy problems and solutions, increase public participation in the areas of concern to sustainable agriculture, and educate the general public about how agriculture is affected by federal policy. Funded mostly by foundations, the National Campaign is a networking organization whose members include family farmers, environmentalists, consumers, and social and economic justice advocates. It holds an annual meeting that includes educational workshops, alliance-building sessions, and short- and long-term planning sessions. The 2002 meeting included 135 people representing 101 organizations. The campaign publishes a quarterly newsletter, *Ag Matters*, and sends frequent policy advisories through its "Action Alert" e-mail list. The National Campaign works closely with one state and four regional Sustainable Agriculture Working Groups: California, Northeast, Midwest, Southern, and Western.

The California Sustainable Agriculture Working Group (CalsAWG), also founded in 1994, is a coalition of California organizations dedicated to building and strengthening the state's movement for a sustainable and socially just food system. CalsAWG provides a forum for information exchange and collaborative action and advocacy. It became incorporated as a nonprofit membership organization in July 2002. Prior to that it had been a project of the Community Alliance with Family Farmers. It now has thirty-five member organizations, including ones that represent farming, farm labor, environment, and public health.

Community Food Security

The movement for community food security is the most recent iteration in approaches to solving food security problems that have existed for millennia. Community food security builds on concepts of previous food security efforts while offering alternatives to what those in the movement see as partial or short-term efforts to solve food security problems. In this section I summarize the history of the concept of food security, discuss the rise of the community food security movement, and describe the development of community food security movement organizations.

Concepts of Food Security

The concept of community food security is of more recent origin than that of sustainable agriculture. As with sustainability, however, concerns about achieving food security are not at all new. While food security issues of one form or another have been with us since the beginning of time, contemporary approaches to food security emerged during the world food crisis of the early 1970s. In 1974 the United Nations convened the World Food Conference in response to unprecedented increases in world prices of staple foods. The issue of food security was the dominant theme of the conference, and food security became a clear and central policy goal of most developing countries (Chisholm and Tyers 1982). While initially food security was usually defined at the national or global scale, it soon became clear that these types of aggregate measures missed conditions of food insecurity within households, communities, and regions and that these arenas needed to be addressed as well.

In the United States, where food insecurity has been a persistent, if less severe, problem than in impoverished countries, concepts of hunger and malnutrition were medicalized prior to the 1980s. Hunger was defined in clinical terms in order to facilitate measurement techniques that would "presumably provide the hard evidence from which to draw conclusions about the incidence of hunger" (Eisinger 1996: 218). During the 1980s, however, it became clear that, for policy purposes, it was more important to define conditions that lead to hunger, since by the time clinical effects of hunger become apparent, the damage may be irreversible (Neuhauser et al. 1995). Food security became the new discourse, defined by the USDA (1998a:1) as a condition in which "all people at all times have access to enough food for a healthy, active life. At a minimum, food security includes the ready availability of nutritionally adequate and safe foods and the assured ability to acquire acceptable foods in socially acceptable ways (for example, without resorting to use of emergency food supplies, scavenging, stealing, and other coping strategies)." Food insecurity is measured by a combination of factors. These include incidence of food acquisition through "abnormal" channels, (e.g., emergency food programs or borrowing from friends); limitations on the variety and quantity of available food; worries about obtaining sufficient food supplies and the money to buy them; and a poor-quality diet (Eisinger 1996).

The amount of political attention paid to hunger in the United States has gone through cycles of official concern and indifference. It was not until the Depression that federal food assistance programs were developed in an attempt to reconcile the paradox of the simultaneous existence of agricultural surpluses and hunger (Poppendieck 1995). These early programs focused primarily on the disposal of agricultural surpluses. Contemporary efforts to end domestic hunger began in the late 1960s when hunger was "discovered" in America in the Mississippi Delta by Senators Robert F. Kennedy and Joseph Clark following President Johnson's 1964 declaration of a war on poverty. At this time, social programs were instituted to more directly combat hunger (Fitchen 1997). These programs included food stamps, school lunches, and supplemental food for women, infants, and children (wic). Even with the emphasis on feeding the hungry, agricultural interests continued to hold sway in the development of the programs. For example, the usda food stamp program was originally developed largely through the self-interested, rent-seeking behavior of economic agents in the food industry rather than social welfare (DeLorme et al. 1992). Nonetheless, these programs made significant improvements in food security for low-income people.

Eventually, however, the slowdown of the postwar economic boom, the breakdown of the political contract between capital and labor, and the increasing influence of conservative elements in government resulted in a new food security crisis. In the 1980s many people's economic conditions worsened; low-income people lost the ground they had gained, and many middle-class families joined the ranks of the newly poor. Children were far from immune to these trends. Between 1989 and 1993 there was a 26 percent increase in the number of children living in families with incomes below 75 percent of the poverty line (Food Research and Action Center 1995). Despite these conditions it was during this time that policymakers began cutting safety-net food programs and the farm surplus–food stamp coalition began to disintegrate. At this point, the emphasis shifted away from government assistance and toward private-sector emergency-food programs to accommodate increasing food needs. The Medford Declaration of 1991, produced by leaders in the antihunger movement, called for voluntary, community-based efforts to provide food as a supplement to public food-assistance programs aimed at achieving food security (Poppendieck 1997). Recognizing the insufficiency of efforts to combat the scope of deteriorations in food security, activists began developing a more articulated community approach to food security.

Development of Community Food Security

In many ways the origins of the contemporary community food security movement can be traced to the uprising in Los Angeles following the Rodney King verdict in 1992.[5] One of the vulnerabilities exposed through the uprising was that of food access and quality in low-income communities. This prompted a group of environmental justice students from the University of California at Los Angeles (UCLA), led by Robert Gottlieb, to undertake a research project on the core issues facing the ethnically diverse and limited-resource community of South Central Los Angeles. The research project resulted in the publication of *Seeds of Change: Strategies for Food Security for the Inner City* in 1993. The needs assessment found that the most pressing concern of people in the community was food—access, quality, and price.

Bolstered by these findings, and building on the long-standing efforts of people working on food issues in local communities, a small group of people decided that it was time to come together to look at food security problems and efforts in a more comprehensive way. In 1994 Bob Gottlieb (then a professor at UCLA), Mark Winne (executive director of the Hartford Food System), and Andy Fisher (then a graduate student at UCLA) convened a group of thirty organizations and individuals in Chicago to discuss new approaches to food security. A key impetus for holding the meeting at this time was the possibility of influencing the upcoming farm bill legislation, which authorizes programs and funding for food and agricul-

---

5. In April 1992 four European-American Los Angeles Police Department officers were acquitted of charges of committing assault in the process of arresting African-American motorist Rodney King. Hours after the verdict was announced an urban civil disturbance was in full swing in the city.

ture in the United States. The experience of attempting to forge effective coalitions of sustainable agriculture, rural development, and antihunger groups to affect farm bill legislation in the previous decade highlighted the need to develop approaches that could address and integrate social, economic, and environmental issues in the agrifood system (Gottlieb 2001). Through the organizing begun in Chicago, an effort emerged during the debates over the 1995 farm bill to identify community food security as the conceptual basis for advancing changes in the food system. As a result, the 1995 Community Food Security Empowerment Act proposal was developed and eventually endorsed by more than 125 organizations, including antihunger and sustainable agriculture groups (Gottlieb and Fisher 1996b). Advocacy of the act led to the inclusion of community food security initiatives in the 1995 farm bill.

The CFSC (1994) defines "food security" as the ability of "all persons [to obtain] at all times a culturally acceptable, nutritionally adequate diet through local non-emergency sources." To achieve this goal, community food security efforts focus on seven areas: community food planning; direct marketing; community gardening and urban food production; strengthening food assistance; farmland protection; food retail strategies; and community and economic development (Fisher and Gottlieb 1995).

Community food security (CFS) has six basic principles (Community Food Security Coalition n.d.). The first is that community food security is focused on meeting the food needs of low-income people. The second is that it addresses a broad range of food-system issues, including farmland loss, agriculture-based pollution, urban and rural community development, and transportation. The third principle is community focus. Community food security works to develop a community's resources so it can meet its own needs through, for example, farmers' markets, improved transportation systems, urban agriculture, and community-based food processing. The fourth principle—self-reliance and empowerment —is closely related to community focus. Community food security emphasizes improving the abilities of individuals to provide for their own food needs; this includes involving community residents in planning, implementing, and evaluating community food security efforts and projects. The fifth principle asserts that a local agricultural base is central to community food security. Local agriculture is valued because it builds stronger ties between consumers and farmers, educates consumers about their food sources, protects farmland from development, and provides better access to markets for farmers, who in turn will be able to pay better wages to farmworkers. The sixth principle holds that CFS is systems-oriented and that it include a wide range of disciplines and value collaborations among multiple and diverse organizations.

Comparing Community Food Security and Traditional Antihunger Approaches

The community food security movement arose out of a desire for more comprehensive approaches to food security. Community food security is simultaneously a goal, an analytical framework, a dynamic concept and strategy for movement building, and a tool for innovative policy development (Gottlieb and Joseph 1997). At the same time, it embodies a critique of traditional antihunger programs. The differences between traditional antihunger and community food security approaches have been summarized by Winne, Joseph, and Fisher (1997), key founders of the community food security movement (Table 3). As opposed to the concept of hunger, which measures an existing condition and is defined in terms of an individual's food insecurity, community food security has come to represent a community-based and prevention-oriented framework. "It seeks to evaluate the existence of resources, both community and personal . . . to provide an individual with adequate, acceptable food" (Gottlieb and Fisher 1996b: 196). While traditional food programs are based on the idea of food entitlements or charity, community food security emphasizes food self-reliance. For the community food security movement, traditional programs such as food stamps and food banks are seen as stopgap measures that fail to address the need for long-term solutions to food security.

Community food security works to build a community-based food system grounded in regional agriculture and local decision making. While the federal government defined food security as "a condition in which all people have access at all times to nutritionally adequate food through normal channels" (House 1989), the community food security movement has an added emphasis on local sources of food. In its definition of food security, the movement substitutes the words "local, non-emergency sources" for "normal channels" to specify acceptable food sources. In contrast, the antihunger movement generally has not focused on how or where food is produced (Winne et al. 1997), and has not viewed as problematic the fact that most major food program decisions are made at the federal rather than local level. Community food security activists are also concerned about the nutritional quality of people's food. Food banks and government commodity programs distribute the surplus from the regular food system, not necessarily food that contributes to a balanced and healthy diet.

Table 3 Comparison of anti-hunger and community food security concepts

	Antihunger	Community Food Security
Model	Treatment; social welfare	community development
Unit of Analysis	Individual/household	Community
Time Frame	Shorter-term	Longer-term
Goals	Social equity	Individual empowerment
Conduit System	Emergency food, federal food programs	Marketplace, self-production, local/regional food
Actors	U.S. Department of Agriculture, Department of Health and Human Services	Community organizations
Agricure Relationship	Commodities; cheap food prices	Support local agriculture; Fair prices for farmers
Policy	Sustain food resources	Community planning

Source: Adapted from Winne et al. 1997.

Organizations in the Community Food Security Movement

After the Chicago meeting in 1994, a loose coalition of organizations continued to work on farm-bill legislation in an attempt to integrate ideas and practices of community food security. Out of this effort grew the national Community Food Security Coalition (cfsc), established in February 1996 (and incorporated in 1997). The cfsc is a nonprofit North American organization working to build sustainable and regional food systems and improve access to nutritious food. As a coalition it consists of groups centered around social and environmental justice, nutrition, environmental protection, sustainable agriculture, community development, labor relations, and antipoverty and antihunger efforts. The cfsc has a committee structure for addressing issues of concern to its members.

The cfsc has developed quickly. By 1998, the organization had grown to more than four hundred members and a mailing list of over four thousand for its newsletter, *Community Food Security News.* Hundreds of projects and conferences have been initiated since 1997. Rapid growth has continued. In 2002 the organization had close to 700 members (265 of which are organizations) and a mailing list of 6,500, with members located throughout forty-one states and the District of Columbia. Its staff has grown from one and a half in 2001 to eleven in 2002 (Fisher 2002a). While the number of participants at annual conferences had been averaging around 300, nearly 600 people participated in its sixth annual CFSC conference in October of 2002.

The CFSC focuses on three primary areas of work: (1) training and technical assistance (e.g., conferences, workshops, community food assessments); (2) project work (e.g., farm-to-school programs); and (3) policy advocacy and organizing at local, state, and federal levels. In addition, the CFSC publishes a quarterly newsletter, policy papers, research reports, and guidebooks; issues policy updates; and maintains a listserve that facilitates information sharing and networking among 500 subscribers. Coalition staff organize about sixty workshops and give about thirty presentations on community food security each year.

After a year of planning and organizing, the California Community Food Security Network was launched in June 2002 at a statewide meeting, "California Community Food Security Summit: Organizing for Action," attended by two hundred members of the California community food security movement. The purpose of the meeting was to take the "first step toward building the cohesion necessary to take the movement to the next stage." It built upon five listening sessions that were held throughout the state to learn more about the food and agriculture issues and priorities of people in diverse communities. While focused on community food security, the conference was sponsored not only by the CFSC but also by thirteen other groups, including food banks, antihunger organizations, environmental groups, sustainable agriculture organizations, and the University of California. Out of the meeting grew the California Community Food Security Network, a consortium that includes organizations representing environmental, nutrition, hunger, farmer, labor, and public health issues. The goal of the network is universal access to healthy food, which is to be achieved through means of education, organizing, and advocacy. Issues addressed include hunger, diet-related health problems such as diabetes and obesity, lack of access to fresh produce, and the loss of family farms. The network intends to develop a coordinated policy platform and improve cooperation among state and local organizations in order to further progress toward community food security.

Both the problems in the agrifood system and the social movements meant to ameliorate them have had a continuing presence in this country.

Problems such as hunger and food safety were recognized and organized against early in the 1800s. Concerns about the structure of agriculture, family farms, and agriculture's effect on the environment go back almost as far. Efforts to change working conditions for agricultural laborers, which began with the abolitionist movement, focused on migrant farm labor starting at the beginning of the 1900s. Movements addressing these various issues have gone through periods of decline and resurgence. Most recently, agrifood issues have been taken up by the movements for sustainable agriculture and community food security. Begun in the last years of the twentieth century, these movements show no sign of ebbing and instead are gaining strength and momentum. Their discourses have proved powerful enough to work their agendas into dominant agrifood institutions such as the USDA and land-grant universities and have also shown sufficient vigor to have created, developed, and maintained new modes of production, distribution, and consumption throughout the country.

3	landscapes of alternative agrifood movements
	institutional integration and construction

If social movements are to be more than ephemeral, they must become part of the fabric that organizes and mediates social relationships. In every society, this fabric is woven out of institutions, which according to Harvey (1996) are human-produced durable spaces that play a key role in the maintenance of the social order. Social institutions comprise the networks that humans have developed over time to establish beliefs and norms, regulate behavior, and promote the "common good." While the content and form of these institutions vary across time, space, and culture, sociologists consider family, education, religion, economy, and government to be the fundamental types. Given that society is made up of social institutions, social movements cannot create social change without either integrating into and reforming existing institutions or creating new forms of institutions in spaces outside the interests and priorities of the ones that already exist. Alternative agrifood movements do both. In some situations these approaches converge, as in those cases where traditional institutions begin to support the development of an alternative institution. This institutionalization of activism typically progresses in stages—from protest to integration and construction.

Reforming traditional institutions is the primary focus of many social movements. For some analysts, in fact, the power of social movements is directly related to how well they are able to engage with and integrate into traditional institutions. In this framework, the success of social movements is measured by the degree to which they are accepted and the extent to which they gain rights to negotiate and consult with existing authorities and established groups (Goldberg 1991). By this gauge, the movements for sustainable agriculture and community food security have been extremely successful. Sustainable agriculture programs have been established in the USDA as well as at many of the largest American agricultural universities, such as the University of California, Ohio State University, and Iowa State. A critical mass of university-sponsored programs in sustainable agriculture was achieved when, in November 2000, nearly thirty representatives of U.S. universities with programs gathered to discuss modes of collaboration and cooperation on sustainable-agriculture issues. The concept of community food security quickly took hold throughout the country and became the foundation of a new federal program only a year after the concept was first formally articulated. This integration of sustainable agriculture and community food security into agrifood institutions reflect the institutionalization of social-movement agendas.

In addition to working within existing institutions, social movements also work to construct new social forms outside existing institutions that can facilitate the achievement of their movement goals. According to some scholars, this latter approach is likely to be the most effective. For example, Kloppenburg and others (1996: 38) write that, in the current agrifood system, "neither people nor institutions are generally willing or prepared to embrace radical change." Rather than challenge the system from within, then, radical reformers advocate "secession" or withdrawal from the dominant food system and the creation of alternatives. For them this involves a gradual "hollowing out" of the global food system by reorganizing "our own social and productive capacities." What parallels such secession is "succession"—forming new relationships that slowly move the old food system to a new one. This gradual approach of developing new agrifood institutions outside the traditional system has been a major thrust of contemporary alternative agrifood movements.

This chapter looks at both of these approaches—working within traditional agrifood institutions as well as creating new ones. First, in order to contextualize the work of alternative agrifood movements within traditional institutions, I briefly recount the framework and history of American agrifood institutions, including the relationship between food and formal politics. I then point to programs that have been developed in sustainable agriculture and community food security that exemplify the integration of alternative agrifood priorities into traditional agrifood institutions. Next, I provide examples of new agrifood institutions that are being developed by the alternative agrifood movements. Last, I explore the impacts and interstices of these changes.

Food as Politics

Food has always been political. The need for food required the creation of social relationships of coordination and cooperation. Increased food production enabled people to create social surpluses, providing resources for the development of society and political institutions. Agricultural policies are among the oldest and most significant areas of state intervention. As early as 4000 B.C., villages began to consolidate into larger political units in order to reap the benefits of large-scale irrigation (Worster 1985). Agriculture plays a special role in the historical and contemporary nation-state, since it has been central to the development and maintenance of society and the state's ability to function. Busch and Lacy (1984) remind us that ancient civilizations rose and fell based upon their ability to maintain a secure, stable food supply. According to Paarlberg (1983), a productive agricultural sector is the only way that governments can satisfy both producers and consumers.

Since the beginning of colonization, the American food and agricultural system has been largely created and closely defined through public policies and government programs. For example, agricultural land was allocated by the English government and laws were passed requiring farmers to plant corn as well as tobacco (Ebeling 1979). Articulated, large-scale federal involvement in the U.S. agrifood system dates back to the latter part of the nineteenth century. At the time, this intervention reflected an anomalous governmental position, since it occurred within a putative period of laissez-faire economics in which government involvement in economic development was disdained (Danbom 1987). Nonetheless, in 1862 the USDA and state agricultural colleges were established, followed in 1887 by the state agricultural experiment stations.[1]

Agriculture is the only sector of the U.S. economy for which it can be said that there is national planning (Shover 1976). The agricultural industry has a federal cabinet-level agency, the USDA. It is the largest cabinet

1. The U.S. Department of Agriculture was created by taking the Bureau of Agriculture out of the Patent Office; it became a cabinet-level agency in 1889 (Danbom 1997).

department after Defense and Treasury (Office of Management and Budget 1995), and it administers more regulatory laws and programs than any other government agency. If the term "planning" seems like an exaggeration, it is nevertheless true that the agrifood system is largely determined through public policies and heavily subsidized by public funds. Because of the pervasive involvement of the government in agricultural policy, both the successes and failures of agriculture have been attributed to the state (de Janvry 1983).

Agricultural planning has been achieved not through regulation and directives so much as by offering various services and opportunities—such as research, extension, farm credit, insurance, disaster assistance, conservation, and commodity subsidies—that farmers may take advantage of or not, as they see fit. Most government involvement in agriculture has occurred through such voluntary programs. Government programs for agriculture have not directly controlled it so much as provided programs from which growers can benefit (Merrigan 1997). Even market regulations are designed to benefit farmers by ensuring high prices for their products, and food quality programs (like meat inspections) protect farmers from market disruptions. In addition to programs like crop subsidies and irrigation projects, public agricultural science has played a significant role in shaping the current conditions of U.S. agriculture. The food and agriculture research system determines how problems are defined, how solutions are derived, which options are considered available, and what types of changes are likely to take place.

Why has the government been so involved in the agrifood system? A number of rationales have been put forward over the years. One is agriculture's greater dependence, relative to other industries, on factors such as the weather that render farm products highly perishable. Another is the price instability of agricultural products; because production commitments are set once the crop has been planted, farmers must sell at whatever price they can get. They cannot adjust their production to the market. Then there is the relatively inelastic demand for agricultural products; that is, there is a more or less fixed amount of food that people buy. Another is the importance of agriculture to the American economy. In most countries the agricultural sector has been a primary arena for capital accumulation. The food industry is the largest in the United States, accounting for 18 percent of jobs and 18 percent of the gross national product (Economic Research Service 1994). It is the second most profitable industry in the United States, surpassed only by the pharmaceutical industry (Magdoff et al. 1998). The agricultural sector is a large contributor to America's export volume and therefore its balance of payments. While the U.S. imports more than it exports overall, this is reversed in agriculture. Agricultural exports also function as a "loss leader" for the development of overseas markets for other U.S. products and industries.

Maximizing production is seen as the way to both increase farmers' profits and maintain a cheap and abundant food supply to benefit consumers and their employers. Food is a major component of the determination of a "living wage" for workers. High agricultural production keeps food prices low, which minimizes labor costs for other sectors of the economy. The inelasticity of food demand, however, is a problem for this assumption and has led to much of the justification for state intervention in the agricultural sector. Historically, overproduction of agricultural commodities has led to the collapse of farm prices and the decrease of farm income.[2] Government policies have therefore also been designed to equalize incomes between farmers and nonfarmers. The rationale for state funding of agricultural research has been that, given the dispersed nature of agriculture, farmers are unable to capture the benefits of research they conduct themselves. In other words, farmer innovators, unlike Monsanto, would not have been able enforce their intellectual property rights in order to capture the economic rents that could accrue from their innovations. In addition, agricultural research has been publicly funded based upon the assumption that since its benefits accrue to society as a whole, society should bear the costs (Hadwiger and Browne 1987).

Institutions such as the USDA and agricultural universities are "hegemonic" institutions in the sense that they develop and transmit ideas, discourses, and practices that constitute the "common sense" of the agrifood system. These institutions have been key partners in the industrialization of the American agrifood system. Can they play a similar role in the development of environmentally sound and socially just agrifood systems? The next section explores the degree to which priorities of sustainable agriculture and community food security are being integrated into these institutions.

The Integration of Alternative Agrifood Priorities into Dominant Agrifood Institutions

Alternative agrifood movements cannot afford to overlook the role of public institutions in the United States. These movements work to open

---

2. Still, farm policies have provided incentives for farmers to expand production even when the market for their products has been saturated.

democratic space within traditional agrifood institutions from which to reshape them. What are the pathways through which change takes place and how successful have the alternatives been? Answering these questions requires examining those institutional spaces where sustainable agriculture and community food security perceptions and practices have been inserted into existing public and nonprofit food programs and agriculture institutions. Of particular importance to such studies are those institutions in California and the United States that have been primary producers and distributors of dominant agrifood-system ideas and practices.

The USDA has established programs in both sustainable agriculture and community food security, and both—the program for Sustainable Agriculture Research and Education (SARE) and the Community Food Project (CFP)—fund competitive grants. Similarly, the University of California has a program—the Sustainable Agriculture Research and Education Program (UC SAREP)—that provides competitive grants focused on sustainable agriculture and community food security. The perspective of a private foundation—America's second largest—is represented by the W. K. Kellogg Foundation's Integrated Farming Systems (IFS) initiative and its California product, the California Alliance for Sustainable Agriculture (CASA).

Each of these programs—SARE, CFP, UC SAREP, and the Kellogg IFS initiative—is embedded in powerful and influential institutions in the American agrifood system. SARE was selected because it is a federally funded, essentially national, program that is insulated from the commodity- and corporate-driven politics of land-grant universities. UC SAREP operates inside the University of California, which is by far the nation's largest land-grant university, with an agricultural research allocation that is nearly 60 percent higher than the next best-funded land-grant institution (NRC 1995). UC SAREP serves a state with an agricultural system and culture that are quite different from those of the Midwestern agricultural "heartland," which is often taken as a baseline for many agricultural programs. Although structurally detached from it, for years the Kellogg program has focused on catalyzing significant changes in the land-grant system.

I focus to a large extent on these programs because they include competitive grants. Grants awarded reflect the interests both of the institutions and those of movement participants who apply for funding. Priorities and pathways for achieving sustainable agriculture and community food security are operationalized primarily through these competitive grant programs. There are three points at which appropriate strategies are framed and selected. First, the request for proposals outlines program priorities, setting the parameters for which subjects and types of projects will be considered for funding. Second, individuals and organizations write and submit proposals. These proposals are reviewed and ranked by panels which then recommend certain projects for funding. Proposals are generally ranked based on problem importance and project feasibility, and in the case of the SARE and SAREP programs, on scientific merit.

Institutions and leaders imbued with expert status have the power to construct and determine social and political "reality." Programs in sustainable agriculture and community food security mark decisive moments of political and ideological construction in which the meanings of these social movements are codified and operationalized and problems are defined, studied, and "solved." These programs are moments of "discursive closure," points at which it is possible to determine which aspects of the problem have been included and which have been ignored (Hajer 1995). Programs in sustainable agriculture and community food security also represent moments of what I would call strategic closure, wherein certain methods and agents of change are retained while others are discarded.

USDA *Sustainable Agriculture Research and Education Program*

Through subtitle C of the Food Security Act of 1985 (the farm bill of that year), Congress authorized the USDA to develop a program in sustainable agriculture. Within this legislation, the Agricultural Productivity Research Act mandated a federal research and education program in sustainable agriculture. The program was authorized for an indefinite period of time (i.e., it does not require reauthorization). According to the U.S. General Accounting Office (1992b), this provision was included in the farm bill because of "concerns about the environment and farmers' dependence on mechanical and chemical inputs." It was first funded by Congress in December 1987 and initiated in 1988. The primary goal of the program was to "develop and promote widespread adoption of more sustainable farming and ranching systems that will meet the food and fiber needs of the present while enhancing the ability of future generations to meet their needs and promoting quality of life for rural people and all of society" (Madden 1998).

Originally called the Low-Input Sustainable Agriculture program (LISA), the name was changed in 1990 to the Sustainable Agriculture Research and Education program (SARE). SARE includes a competitive grants program, a professional development program for training agricultural field personnel (mainly land-grant agricultural extension agents) in sustainable agriculture, and an information dissemination arm called the Sustainable Agriculture Network (SAN).

Headquartered in Washington, D.C., SARE is directed by four independent regional administrative councils—the Northeastern, Southern, North Central, and Western. Membership on administrative councils is decided by participants in the region represented. They typically include farmers, extension agents, researchers, businesspeople, and representatives from state, federal, and nongovernmental organizations. The regional councils provide policy direction for the program, identify information needs, and choose projects for funding after a technical review panel has ranked them. The purpose of reviewing grants at the local level is to ensure that priorities are set by people who live in the area (SARE 1998a).

SARE has funded approximately eighteen hundred projects in both research and education and professional development since its inception as LISA in 1988.[3] The regional councils decide how the funds are to be allocated. Projects funded are usually interdisciplinary partnerships of scientists, producers, and others and include research on crop trials, marketing, quality of life, integrated farming systems, and resource conservation. SARE also funds research and demonstration efforts such as the development of farmer-to-farmer networks.

SARE's Professional Development Program, which was authorized in the 1990 farm bill, provides funds for planning and grants to develop education and outreach programs and strategies in sustainable agriculture. Programs are designed to work with Cooperative Extension agents, resource agency personnel, and others who work directly with farmers and ranchers.

SAN facilitates information exchange through printed and electronic communications on practical questions about sustainable agriculture. Established in 1991, its programs are intended to provide information related to SARE programs and objectives and to develop innovative approaches to communication for the SARE program. SAN operates as a cooperative effort of university, government, farm, business, and nonprofit organizations. SAN publishes primarily practical information on sustainable agriculture through both printed and electronic media. Its publications include titles such as *Managing Cover Crops Profitably* and *Steel in the Field: A Farmers' Guide to Weed Management Tools*. SAN also moderates an elec-

---

3. In 1991 a new program funded jointly by the USDA and the Environmental Protection Agency was established to focus specifically on environmental issues.

tronic sustainable-agriculture discussion group (sanet-mg) that has more than 750 subscribers. As part of its networking functions, SAN also maintains a "Calendar of Sustainable Agriculture Events" on its web site to inform people about sustainable agriculture activities in their area. SAN works both with SARE participants in each of the four regions and with grassroots sustainable-agriculture organizations.

USDA *Community Food Projects Program*

In 1996, Congress authorized a program of federal grants to support the development of community food projects as section 25 of the Food Stamp Act of 1977. In this legislation, community food security is defined as "all persons obtaining at all times an affordable, nutritious and culturally appropriate diet through local, non-emergency food sources (or through normal economic channels)" (House 1995). The goal of food security programs is to "build economically strong local food economies in which individuals can achieve long-term self sufficiency."

The program is administered through the USDA's Cooperative Research, Education, and Extension Service and operates as a competitive grants program. Priorities established for the grants are designed to encourage programs that address the food needs of low-income people; that increase the self-reliance of communities in providing for their own food needs; and that promote comprehensive responses to local food, farm, and nutrition issues (USDA n.d.).When the program was established, Congress authorized $1 million for fiscal year 1996 and $2.5 million for each of the fiscal years 1997 through 2002. In the 2002 farm bill the funding was doubled to $5 million per year of mandatory funding through 2007 ($30 million total). Federal funds received by grantees require matching amounts, which can be in-kind contributions. Grants are intended to provide one-time seed funding to develop projects that can then become self-sustaining. A group appointed by the USDA in consultation with nongovernmental organizations reviews proposals. Since 1996, forty-nine grants have been awarded to community-based organizations in every region of the country (USDA n.d.). Funded projects have included those focused on job training, employment opportunities, small business expansion, neighborhood revitalization, open-space development, transportation assistance, and other community enhancements (USDA 2002).

The Community Food Projects program specifies that, in order to receive funding, proposed projects should be whole-systems oriented and include community development, economic opportunity, and environmental enhancement among their priorities. Comprehensive solutions may include elements such as improved access to high quality, affordable food among low-income households; expanded economic opportunities for community residents through local businesses or other development programs, improved employment opportunities, and job-training, youth-apprenticeship, school-to-work transition, and similar programs; and support for local food systems, from urban gardening to local farms that provide high-quality fresh foods, ideally with minimal adverse environmental impact.

University of California Sustainable Agriculture Research and Education Program

The University of California Sustainable Agriculture Research and Education Program (SAREP), headquartered at the University of California at Davis, was established about the same time as the SARE program. Its mission is to provide "leadership and support for scientific research and education that promotes agricultural and food systems that are economically viable, sustain natural resources and biodiversity, and enhance the quality of life in the state's diverse communities" (UC SAREP 2000). SAREP was created through the efforts of grassroots organizations working with legislators. The resulting Sustainable Agriculture Research and Education Act of 1986 was passed in response to the "growing movement in California and the nation to change farming techniques by adopting more resource-conserving, energy-efficient systems of agriculture" (California Food and Agriculture Code 1986). The purpose of the legislation was to promote more research and education on sustainable agricultural practices, such as organic methods and biological control of pests, as well as to analyze economic factors affecting the long-term sustainability of California agriculture. The bill directed the Regents of the University of California to establish the Sustainable Agriculture Research and Education Program to (1) develop a competitive grants program; (2) provide information through demonstrations, publications, and other means; and (3) establish long-term research projects on sustainable agricultural practices and farming systems on university-owned research land. In 1994 the California legislature established the Agricultural Chemical Reduction Pilot Demonstrations Projects (Statute AB3383), directing the University of California to develop a program to demonstrate and expand the use of biologically integrated farming systems that reduce use of farm chemicals (California Food and Agriculture Code 1994). The university chose UC SAREP to implement the legislation, which directed that the projects should be selected through a competitive grants process. This became the Biologically Integrated Farming Systems (BIFS) program.

As specified by the 1986 legislation, a Program Advisory Committee and a Technical Advisory Committee provide guidance to SAREP. The Program Advisory Committee recommends goals and priorities for the program and reviews grant applications to determine if they meet program goals. The Technical Advisory Committee evaluates the scientific merit of grant applications. A separate Program Advisory Review Board assesses the proposals and projects of the BIFS program that funds demonstrations of integrated farming systems. In 1999 SAREP also developed a special grants program focused on developing alternatives to methyl bromide, a soil fumigant scheduled to be banned in 2005.

Between 1987 and 2002 SAREP granted more than $7 million to 302 research, education, and demonstration projects. SAREP's constituency includes farmers, farmworkers, educators, policymakers, consumers, and community organizations in California. In addition to managing the competitive grants program, SAREP staff conduct research and education projects on sustainable agriculture and community food systems. Currently, SAREP's three major themes are ensuring the long-term viability of California agriculture; partnering with farmers to implement biologically integrated farming practices; and linking farmers, consumers, and communities through sustainable community development and public policy (UC SAREP 2000).

The first of these themes—ensuring California agriculture's long-term viability—focuses on improving sustainability and environmental quality on farms. Projects include soil management, environmentally safe pest management methods, crop diversification, and organic farming methods.

The second theme—partnering with farmers to implement biologically integrated farming practices—is designed to demonstrate and expand the use of integrated farming systems that economically reduce the use of agrichemicals. Focusing on demonstration projects and education efforts, these projects include on-farm demonstrations of biologically based farming systems, assisting farmer decision making by monitoring biological and environmental variables, and developing public-private partnerships for sharing information about farming systems. BIFS projects have been undertaken in rice, walnuts, citrus, prunes, strawberries, apples, wine grapes, and dairy/forage crops.

The third area—linking farmers, consumers, and communities through sustainable community development and public policy—includes topics that span the whole of the agrifood system. This Community Development and

Public Policy section of SAREP emphasizes the development of community food systems. SAREP defines a community food system as "a collaborative effort to promote sustainable food production, processing, distribution and consumption in order to enhance the environmental, economic, and social health of a particular place" (UC SAREP n.d.). Project goals include improving access to an affordable and nutritious diet, improving direct links between farmers and consumers, developing community-based food-related businesses, improving working and living conditions for workers in the agrifood system, and creating public policies to support sustainable food systems.

Kellogg's IFS Initiative

The W. K. Kellogg Foundation was established in 1930 to help people improve their quality of life and that of future generations. One of the W. K. Kellogg Foundation's eight major programmatic areas is its Food Systems and Rural Development program, which encompasses a number of programming initiatives related to improving various aspects of the agrifood system. Of its five initiatives in this area, the one on which I primarily focus is its IFS initiative. Other agrifood-related Kellogg initiatives include its Food Systems and Rural Development program and its Food and Society program. The W. K. Kellogg Foundation also funds the Consortium on Sustainable Agriculture Research and Education (CSARE), which has been focused on creating change in public agricultural institutions and policies.

The IFS initiative was designed to improve the functioning of America's food system to the point that it provides access to safe and nutritious food for all segments of society. For the Kellogg Foundation, this includes environmental protection, building partnerships between producers and consumers, developing better connections between institutions of higher education and communities, and making relevant marketing and policy changes (W. K. Kellogg Foundation n.d.-a). The IFS program, started in 1993 and completed in 2003, had two phases. In Phase 1, the Foundation committed approximately $16 million to fund eighteen community-based projects throughout the country. It also provided funds for activities and meetings to link the projects together (W. K. Kellogg Foundation 1994). The goals of the IFS initiative were

> To help farmers adopt more integrated and resource-efficient farming systems that maintain agricultural produc-

tivity and profitability while protecting the environment and personal health; and to assist farmers and others in rural communities to empower themselves to address the barriers associated with adopting more resource-efficient and integrated systems so that these systems could become a foundation on which to revitalize and rebuild the economic and community bases of rural America. (Hesterman and Thorburn 1994)

Projects were designed to develop, test, and implement sustainable agricultural practices, and collaboration was a major focus. All projects needed to be organizational collaborations that included an educational or science-based institution. Priority was given to projects that brought together farm and nonfarm leaders to explore ways to resolve conflicts and create a shared vision of a more healthful food and agriculture system. The Kellogg Foundation also gave funds, primarily to NGOs, that helped to increase NGO credibility and that strengthened their position with respect to traditional agricultural institution partners, such as land-grant colleges (Fisk et al. 1998). In Phase 2, begun in 1996, the Foundation funded nineteen projects that continued the efforts initiated in the first phase but which were broader in scope. Projects funded included a community food resource center, an urban agriculture project in a low-income neighborhood, and a farmland preservation project (W. K. Kellogg Foundation n.d.-b).

The California Alliance for Sustainable Agriculture

One of eighteen projects involved in Kellogg's Phase 1 IFS initiative was the California Alliance for Sustainable Agriculture (CASA). At the Kellogg Foundation's request, the project was developed as a collaborative activity among several organizations that had initially applied separately (or with different collaborators) to the foundation's initiative. Thus, an experiment in blending organizational priorities and cultures was set in motion. Original collaborating organizations included the Bio-Integral Resource Center; the California Action Network, which became the Community Alliance with Family Farmers Foundation; the Lodi-Woodbridge Winegrape Commission; the Rural Development Center; the California Institute for Rural Studies; the University of California Division of Agriculture and Natural Resources; the University of California at Santa Cruz Center for Agroecology and Sustainable Food Systems; and the University of California Sustainable Agriculture Research and Education Program (UC

SAREP). These groups brought together long histories of leadership in sustainable agriculture at different yet complementary levels. Each organization was expected to contribute to CASA's overall objectives in those areas in which they had the greatest relative strength and experience, with the idea that this combined effort would prove far more effective than what any single organization could accomplish on its own. Active from 1993 to 1997, the CASA project, "Creating Sustainability in California: Reconnecting People and Environment in Food and Agriculture," had as its purpose to "enhance sustainable food and agriculture systems by creating innovative models for community-based education and coalition building." Target audiences included farmers, consumers, farmworkers, policymakers, and students ranging from elementary school to university levels.

Challenging and making changes within traditional agrifood institutions such as the USDA and the University of California has been a key strategy of alternative agrifood movements. Such efforts have ensured that these institutions begin to pay attention to issues of sustenance and sustainability.

Development of New Agrifood Institutions

This "top-down" effort to create reforms within traditional agrifood institutions is complemented at local levels by "bottom-up" efforts to create new, alternative institutions that can serve as the basis for rebuilding the agrifood system in ways that are more environmentally sound and socially just. In this "consumption politics of food," activism often takes the form of organizing efforts grounded in civil society (Buttel 2000). Farmers and consumers previously marginal to the dominant food system have taken up the unoccupied spaces in this system to develop alternative production, marketing, and social movements around food issues (Whatmore 1995). These agrifood activities and practices engage the imaginations, hopes, and energies of people located in very different sites within the agrifood system.

While collectivities of agrifood system alternatives are referred to by various terminologies, there remains a certain consistency in the types of alternative agrifood institutions named. Among the terms used to designate alternative agrifood systems are the following: "alternative food regimes" (Friedmann 1993); "alternative food systems" (Gottlieb and Fisher 1996a); "local food systems" (Henderson 1998); "integrated food systems" (Clancy 1997); "sustainable food systems" (Pretty 1998); "alternative food streams" (Grey 2000); "alternative food networks" (Marsden 2000); and "alternative geographies of food" (Whatmore and Thorne 1997). Within these various frameworks, the alternative agrifood institutions include farmers' markets, urban agriculture projects, community gardens, community-supported agriculture, food policy councils, school gardens, food cooperatives, and food-based education (Table 4).

What is the value of this "constructivist" strategy of creating new and alternative institutions? First, it is practical. The U.S. Conference of Mayors (1985) suggested that cities could work to improve the health and welfare of low-income citizens. The mayors called for, among other things, increased household food production and preservation, expanded community gardens, improved transportation systems to facilitate food access, new farmers' markets, and the preservation of farmland around urban areas. As framed by Gottlieb and Fisher (1998), community-based production and distribution are seen as creating "new economic spaces" that establish alternatives to the transnational and corporate food system. The movements for sustainable agriculture and community food security both create new modes of food production and distribution and reinvigorate old forms, such as farmers' markets and community gardens. Transitioning to a better food system will only be possible if there are practical alternatives to the types of institutions and practices that have created the current agrifood system.

Second, through the process of creating these alternatives, people are able to engage with things that resonate with their daily lives. The work of developing alternative practices and institutions is a real and immediate point of engagement for people as they go about their daily lives. Buttel (2000) suggests that activities like community-supported agriculture and local food-system projects are the primary ways in which consumers are now expressing resistance to problems in the food system. Food-system alternatives can create and connect economic and social spaces and establish new models that engage public concerns about community, social justice, and environmental sustainability (Gottlieb and Fisher 1998).

Third, alternative agrifood institutions involve the cooperation of diverse groups of people and help to establish networks among people who may have not previously fully participated in the system. These efforts place even greater emphasis on the role of consumers in agrifood system change. Some have argued that they provide openings for restructuring and transforming the agrifood system through the agency of consumers demanding safer, quality, natural, and local foods (Murdoch et al. 2000; Goodman 1999; Whatmore and Thorne 1997; Nygard and Storstad 1998). These efforts are also seen as

Table 4 Core forms of alternative agrifood institutions

Author	Collective term	Activities included
DeLind 1994	locally responsive food systems	CSAs, cooperatives, urban gardens, farmers' markets, community land trusts, food policy councils
Clancy 1997	integrated food systems	Farmers' markets, CSAs, labeling, direct marketing, community gardens, value-added marketing, cooperatives
Feenstra 1997	local food systems	Food policy councils, farmer's markets, CSAs, community and school garden, urban farms, college-level educational farms, cooperative agricultural marketing programs
Pretty 1998	sustainable food systems	Direct marketing, community gardens and cooperatives, alternative knowledge networks, eco-labeling
Grey 2000	alternative food streams	Direct marketing, community supported agriculture, food cooperatives
Lacy 2000	local food systems	Farmers' markets, farm stands, CSA, community gardens, sustainable agriculture organizations, community food security coalitions, food policy councils, producer and consumer cooperatives

Source: Allen et al. 2003.

possessing the advantage of connecting farmers and consumers, thereby improving consumer knowledge about the agrifood system.

A clear theme of alternative agrifood movements is the promotion of local food systems. Local food systems may solve many of the problems that concern alternative agrifood system advocates. They are considered to have environmental benefits, such as reducing energy use; social benefits such as creating new opportunities for solving problems of hunger and homelessness; and economic benefits such as improving opportunities for employment (Dahlberg 1994b). Thus they tie together the priorities of the sustainable agriculture and community food security movements. Among the most prominent and frequently cited types of alternative agrifood institutions in local food systems are farmers' markets, community-supported agriculture, institutional purchasing, urban agriculture, and food policy councils, each of which I highlight in this section. The first three, which involve direct marketing, combine the objectives of providing market opportunities for small farmers and increasing consumer access to fresh produce. Urban agriculture is food production within an urban environment. And food policy councils are efforts to bring diverse stakeholders together to focus on policy issues related to food within a local municipality.

Farmers' Markets

Farmers' markets serve the needs of both farmers and consumers by providing a market outlet for small farmers and by increasing consumer access to fresh produce. In the 1970s farmers markets were organized in urban low-income communities to help provide nutritious food to the urban poor. They have provided accessible markets for producers outside the mass market and filled an important niche for consumers who "valued quality and variety over quantity and uniformity or who wished to support local agriculture" (Lyson et al. 1995: 108). Shoppers purchase their food from the farmers who produced it, generally fresh produce but sometimes also processed foods such as cheese or honey. Consumers have reported that their primary reasons for going to farmers' markets is fresh food and direct contact with farmers, reflecting a desire to both improve their diets and reconnect with their food sources. Farmers retain a greater share of the food dollar and often establish personal relationships with those who eat their products. In addition, farmers' markets provide a major marketing outlet for farmers who use environmentally sensitive production methods.

Community-Supported Agriculture

Community-supported agriculture is another approach that can provide small farmers with a market while increasing access to fresh produce. In such a program a group of consumers (shareholders or members) purchase shares at the beginning of the season with the idea that they will receive a portion of the crops produced that year. Consumers pay a fee to a grower and expect to receive in return a weekly share of fresh produce, usually harvested the same day. In many cases, consumers travel to the farm to pick up their weekly box of produce; in others farmers may deliver the boxes to a pickup location in the community. Consumers get a broader selection of fresher produce, and the farmer has a ready market and cash flow. The vision is that farmers and nonfarmers will work together to support each other and build strong community-based economies (Lawrence 1997). One goal of community-supported agriculture is to connect people in an immediate way with farming by, for example, encouraging active involvement of shareholders in farm work (Clunies-Ross and Hildyard 1992). Almost all community-supported farms use organic production methods. Overall, the idea of community-supported agriculture is to "connect local farmers with local consumers; develop a regional food supply and strong local economy; maintain a sense of community; encourage land stewardship; and honor the knowledge and experience of growers and producers working with small to medium farms" (University of Massachusetts Extension 1997).

Institutional Purchasing

Institutional purchasing involves linking local farmers with public institutions that purchase large volumes of food, such as colleges and schools. Such linkages provide significant market outlets for growers. For example, the school food services market alone is estimated at $16 billion per year. Alternative agrifood advocates have focused their efforts primarily on farm-to-school programs. Instigated by farmers, schools, parents, and community groups, these programs are intended to address two problems concurrently: childhood nutrition problems such as obesity and the lack of access to markets for small and medium-size farms. The farm-to-school initiative joins school food services with local farmers in a partnership that is intended to bring fresher, healthier produce to school meals programs while at the same time supporting local farmers by providing an additional source of income and a relatively secure market. Programs vary across the country and are tailored to the needs of school food-service providers and to the extent to which they can or want to be involved. Farm-to-school programs may include salad bars with farm-fresh fruits and vegetables purchased at the local farmers market. In one case a cooperative of small farmers sells produce directly to the local school district.

Farm-to-college projects are also being initiated, with the same goals as farm-to-school programs. In Wisconsin, six college campuses are buying their food for their dining services from local farms, four of which use organic and sustainable production practices. These programs may have even more influence in improving the nutrition and food habits of diners because college dining halls serve three meals a day. Some colleges also feature food from local farms at catered university events.

Urban Agriculture

Another alternative agrifood institution that is gaining ground is urban agriculture, either as a form of self-provisioning or as production for market. Urban agriculture is food production within a metropolitan area; food may be produced on residential plots, public or vacant private land, balconies, or rooftops. New coalitions are coming together to promote various forms of urban agriculture in the United States. These groups may include health and nutrition advocates, university extension agents, emergency food distributors, environmentalists, and community development organizers (Brown 2002).

One-seventh of the world's food supply is grown in cities by 800 million urban farmers; in the United States, 30 percent of agricultural products are grown within metropolitan areas (Smit et al. 1996). The fruits and vegetables that can be grown on small urban plots are an important source of nutrients crucial to overall health. While fruits and vegetables provide only 8 percent of food energy in an American diet, they are a primary source of vitamin C and carotene (USDA 1998a).

The CFSC identifies three primary and overlapping locations and purposes of urban agriculture in the United States: backyard gardens, community gardens, and small or medium-sized commercial farms in and around cities (Brown 2002). Backyard gardeners are those who produce crops on land near their homes. The food they produce is for their own use and that of family and friends, rather than for commercial purposes. These gardens add variety, freshness, and beauty to people's food supply and help to reduce food costs.

Community gardeners grow food on small plots generally owned and operated by a public or private institution such as a city or land trust. The benefits of community gardens include those of backyard gardens; in addition, the crops may be sold or donated to organizations serving the hungry. The community gardening movement has a long history in the United States. The first organized city gardening program was formed in Detroit in 1893 when the mayor made public land available so that people suffering economic hardship could grow some of their own food (Patel 1992). Urban land has been used for charity gardens for the poor and victory gardens to provide food during wartime (the Liberty Gardens of World War I and the National Victory Gardens of World War II), but most were abandoned following World War II (Hynes 1996). The contemporary community gardening effort was started by public housing authorities and was supported by a USDA urban gardening program. Initiated in 1976, it has helped low-income people in cities to grow and preserve vegetables (Hynes 1996). The amount of food that can be produced in U.S. gardens is substantial, comparable in volume to that of the $18 billion-annual U.S. corn crop (Dahlberg 1994b). In addition to providing access to fresh, nutritious fruits and vegetables, community gardening also provides sites for socializing and community organizing. Often community gardens have turned blighted abandoned spaces into lush gardens that offer relief from harsh inner-city conditions. They can also provide safe spaces and arenas for multigenerational and multicultural interactions. They have been used as public spaces for memorials, weddings, potlucks, and classes.

Increasingly, small plots within cities are being reclaimed for market production and food-based microenterprise development. These spaces fit into the third category of urban agriculture: commercial growers who produce food for sale in and around urban areas. Urban farms produce significant amounts of food in the United States. One-third of U.S. farms are in metropolitan areas, and these farms account for 16 percent of farmland and produce 25 percent of crop and livestock sales (Economic Research Service 1993). In Cuba, where 26,000 hectares are cultivated within cities, urban agriculture is credited with playing a big part in Cuba's recovery from the food crisis brought on by the collapse of the Soviet Union and the U.S. embargo (Rosset 1996). Since they connect consumers with the growers, urban farms are viewed by alternative agrifood-system advocates as vital to the development of local food systems.

Food Policy Councils

One of the newest alternative agrifood institutions is the local food-policy council or coalition (FPC). Food-policy councils have been developed in cities across the country as part of comprehensive efforts to reduce hunger and increase food security in some regions. These organizations work to increase the visibility of food issues in a community and may recommend food-related policies to local governments. Canada's Toronto Food Policy Council, for example, has developed a food policy for the city that establishes the right of all residents to adequate, nutritious food and promotes food production and distribution systems that are equitable, nutritionally excellent, and environmentally sound. This Council frames food security as a health issue in which hunger and poverty are viewed as part of the larger health issue, a perspective which sees access to food as not only equitable but economical.

FPCS are usually public/private partnerships that include representatives of multiple sectors of a region's food system. The membership of these councils typically includes representatives from farming, hunger-prevention, retail-food, nutrition-education, food processing, sustainable-agriculture, religious, health, government and environmental organizations. The first FPC emerged in Knoxville, Tennessee, in 1981, and there are presently fifteen food policy councils throughout the United States and Canada. Similar organizations have been established in Connecticut, Tennessee, California, Minnesota, Wisconsin, and Texas.

Impacts and Interstices of Institutional Change

Progress is being made in both modes of institutional change: the development of alternative agrifood institutions, and the integration of community food security and sustainable agriculture into existing agrifood institutions. In addition, cross-fertilization and cooperation is increasing in the spaces between traditional and alternative agrifood institutions. For example, the USDA SARE program uses the mailing lists of nonprofit organizations advocating sustainable agriculture to publicize calls for proposals and disseminate information about sustainable agriculture (Dyer 1999).

Consumers and farmers are working with others in their communities to build alternative agrifood institutions, an effort that is reflected in the growth of these institutions. For example, farmers' markets have expanded quickly in recent years. The number of farmers' markets in the United States grew 63 percent between 1994 and 2000, with 2,863 markets operating in 2002 (Kantor 2001). They account for an estimated $1.1 billion contribution to farmers' incomes (USDA 1998b). Community-supported agriculture is in a phase of rapid development. While there are still few CSAS—only four states have more than thirty—their numbers are growing. The first community-supported farm in the United States was established in 1985; only ten years later there were over five hundred groups in the country practicing community-supported agriculture, and each year the number of participating farms and members increases (Van En 1995). Interest in community gardens has also increased dramatically in the past decade, possibly because of a resurgence of interest in gardening by those concerned with the environment, health, and self-sufficiency. The American Community Gardening Association estimates that there are now more than 150,000 community gardens in the United States, with 30 percent started since 1991. Farm-to-school programs, virtually unheard of a few years ago, have become all the rage in school districts across the country. These efforts to build alternative agrifood institutions are increasingly supported by programs within traditional agrifood institutions. For example, the W. K. Kellogg Foundation sponsored a "Local Food and Farming Conference" in 2001 as the next step in its decade-long effort to promote sustainable food and agriculture systems.

The USDA has established two programs designed to increase access by low-income people to alternative agrifood institutions. In 1992 Congress established the WIC Farmers' Market Nutrition Program (FMNP). The purpose of the program is to provide fresh, nutritious food and to expand the awareness and use of farmers' markets. WIC (Special Supplemental Nutrition Program for Women, Infants and Children) provides additional food, health-care referrals, and nutrition education to low-income women and children up to five years old. The program, which is authorized in thirty-five states (participation is not mandatory), provides coupons that can be used to purchase food at farmers' markets. Federal funds support 70 percent of the program cost, with the remainder borne by the state. In 2000 almost 2 million people received WIC farmers' market coupons, resulting in sales of approximately $17.5 million. Ten years later, the USDA developed a similar program for seniors. In 2002 Congress appropriated $10 million out of the USDA's Commodity Assistance Programs budget for a Seniors Farmer's Market Nutrition Program (SFMNP). The program provides low-income seniors with coupons they can redeem to purchase fresh fruits and vegetables at farmers' markets and CSAs. The purposes of the program are both to provide low-income seniors with increased access to fresh fruits and vegetables and to aid in the development of additional direct-marketing outlets.

The USDA publishes a national Farmers' Market Directory, holds weekly farmers' markets on USDA property, and helps to coordinate farmers' markets at other federal agencies in the Washington, D.C., area. In 1998 the USDA established a toll-free telephone number to provide farmers and consumers with information on these markets. In addition, since all states are required to use an Electronic Benefit Transfer (EBT) system for food stamps as of October 2002, the USDA has developed pilot projects to facilitate the use of the EBT system (which works like debit cards) at farmers' markets. The USDA also maintains a home gardening web site.

In 1997 the USDA established the Small Farms / School Meals Initiative to encourage the development of farm-to-school programs. The U.S. Department of Defense operates a national program to purchase and deliver fresh produce to military installations, federal prisons, and veterans' hospitals, and it added schools in 1994 (USDA 2000). The school portion of the program began delivery to eight schools in 1994, and only two years later had expanded to thirty-two states. Under its Initiative for Future Agriculture and Food Systems (IFAFS), the USDA provided a $2 million grant to fund a consortium of universities, school districts, and nonprofit organizations to develop new farm-to-school programs in California, New Jersey, and New York.

This trend has been accompanied by substantial efforts to develop sustainable agriculture and community food security programs within traditional agrifood institutions. The extent to which sustainable agriculture and food security have been integrated into U.S. food and agricultural institutions in many ways constitutes an enormous achievement. In some respects, it is remarkable that programs in sustainable agriculture and community food security exist at all. None of the present governmental programs in these areas emerged from within their parent institutions. They were instead established by legislative mandate and introduced into mostly inhospitable institutional environments. Neither the USDA nor the University of California actively sought out sustainable agriculture programs. The Community Food Projects program was established through creative congressional negotiations led by key House and Senate staff rather than a decision within USDA to initiate such a program.

Perhaps most remarkable are the changes that have occurred within the USDA itself. Recall the hostility with which alternative agriculture was met not so very long ago. Until the mid-1980s, the idea of environmentalist agriculture was "relentlessly ridiculed" by established agricultural institutions (Buttel and Gillespie 1988). For example, the USDA's study of organic farming in the United States, commissioned by the Carter administration (USDA 1980), was met with intense opposition from the agricultural scientific, policy, and industry communities, even though the report did not completely rule out the use of synthetic chemicals (Youngberg et al. 1993). In 1981 this report was rejected by the new Reagan administration, and the USDA position of coordinator for organic farming was abolished within the first year of the new administration.[4] In 1982, two federal efforts to legislate organic or low-input research and education programs (the Weaver and Leahy bills) failed to pass despite the use of strategically employed, politically neutral euphemisms in the titles of the initiatives.

---

4. The coordinator, Garth Youngberg, went on to form one of the most prominent sustainable agriculture NGOs, the Institute for Alternative Agriculture.

Although Congress authorized a sustainable agriculture program in 1985, the USDA made no requests to fund the program. Finally, in 1988 Congress itself authorized the necessary funding. Even after the USDA launched the program, support for sustainable agriculture within the agency was lukewarm at best. When the USDA sent a press release announcing the sustainable agriculture initiative, it received a critical letter from the Fertilizer Institute. The USDA then apologized for the press release, prepared a letter clarifying statements about soil fertility and groundwater contamination, and allowed Fertilizer Institute officials to review the letter before it was released (House 1988). Congress reprimanded the USDA for its foot-dragging and the degree to which it was beholden to agribusiness interests, stating: "Given the potential environmental and economic benefits of a low input program, it is unfortunate that USDA subjugates its enthusiasm for such a program to concerns about the possibility of raising the 'hackles' of the agricultural chemical manufacturers" (House 1988: 27).

Not until 1991 did the USDA itself request funding for the program, and even then USDA's request for funds fell far below the amounts that had been appropriated by Congress. In fiscal years 1988 through 1992, the USDA requested only a third of the funding appropriated by Congress (GAO 1992b). A congressional report on the implementation of sustainable agriculture programs concluded that not only was the USDA reluctant to develop a national sustainable agriculture program but that the USDA *itself* presented "perhaps the most difficult of all barriers to overcome" (House 1988).

Yet change has come, if gradually. The SARE program is now firmly established. SARE funding has continued to grow since it was first appropriated in 1988. When the federal sustainable agriculture program was first funded in 1988, the allocation represented less than one-half of one percent of the USDA's research budget in that year (calculated from Office of Technology Assessment 1995b). Funding levels for fiscal year 1998 ($12 million) were three times higher than in its first year ($3.9 million) (Madden 1998). The SARE program has continued to receive substantial increases; and according to SARE director Jill Auburn, the program is now "solidly established in the President's budget." While in 1992 the SARE program awarded only thirty grants per year, it now funds two hundred per year (SARE 1998a).

Once initiated, the SARE program began to play a key role in increasing interest in and acceptance of sustainable agriculture among farmers and agricultural institutions (GAO 1992b). For example, the USDA included a chapter on sustainable agriculture in the 1991 *Yearbook of Agriculture* (USDA 1991). The USDA National Agricultural Library established a special Alternative Farming Systems Information Center that produces numerous publications and bibliographies related to diverse aspects of alternative agrifood systems. In 1993, the USDA, pointing to increased international competition and changing consumer preferences, called for a more balanced and sustainable American agricultural system (USDA 1993a). The 2002 farm bill, the Farm Security and Rural Investment Act of 2002, for the first time included research and technical assistance programs designed to assist organic farmers with production and marketing. These type of changes represent a major reorientation away from the dismissiveness and contempt with which terms such as "organic farming" had been treated by the agricultural establishment throughout most of the twentieth century. Indeed, SARE director Dr. Jill Auburn (2002) reports that the SARE program is promoted by the USDA as an example of a successful program that is making a difference in rural communities.

Although the concept of community food security has a much shorter history than that of sustainability, it has also been embraced by the USDA. Unlike the federal sustainable agriculture program, which was established as the result of a long, diffuse effort, the enactment of community food security legislation can be traced directly to the actions of a coalition of activists and academics over a very short period of time. The first meeting to focus on the legislation was convened in August 1994, and the Community Food Projects (CFP) program was established by 1996 as part of the nutrition assistance title of the farm bill. In 2002 the budget of the program was doubled, which is remarkable, given the climate of budget austerity prevailing at the time. Like SARE, the CFP program is well regarded by USDA officials. According to CFP director Dr. Liz Tuckermanty (2002), it is considered by those in the department to be a "small program with a great presence." It serves the direct needs of people in an effective way at a relatively low cost. Since CFP supporters and constituents are generous in their commendations to USDA on the CFP program, the program has become well known in the upper echelons of the organization. Secretary of Agriculture Anne Veneman (known to support large-scale export agriculture) put out a press release on the Community Food Projects program on World Food Day.

In addition to establishing the Community Food Projects program, former Secretary of Agriculture Dan Glickman wanted to create an Office of Community Food Security as part of the legacy of his administration. The

USDA established the Community Food Security Initiative in 1999 to create partnerships between the USDA and local communities. The proposed budget for fiscal year 2000 included $800,000 to provide technical assistance to communities and $15 million for a competitive grants program for food recovery and crop gleaning. This effort indicated the degree to which the interest in institutionalizing community food security within the Department had grown. The executive director of the CFSC wrote that although it might seem an unusual step to take for activists accustomed to protesting government food and agriculture programs, it was time to write a thank-you letter to the secretary of agriculture (Fisher 1998). Another sign of increasing cooperation between the USDA and community food security advocates occurred when the CFSC established leadership awards in 2002, and it gave one of the three awards to Liz Tuckermanty, director of the USDA Community Food Projects program.

The efforts of sustainable agriculture and community food security proponents to work with agrifood institutions have been quite successful. Their initial struggle to get their issues addressed rather than summarily dismissed by these institutions has been effective. The development of programs in sustainable agriculture and community food security movements is especially remarkable given the conservative nature of the institutions in which they have been inserted and the entrenched interests that have dominated these institutions. It is true that programs in sustainable agriculture and community food security remain comparatively small within the scope of their parent institutions, but their impact belies their size, both within their parent institutions and for the people they serve. Programs that have been developed in sustainable agriculture and community food security are not simple accommodations or attempts to defuse or legitimate alternative ideas; they are sincere efforts by those involved in them to solve sustainability and food security problems. It is a testament to their commitment and creativity that they have managed to accomplish so much operating within the institutional framework that developed and generally continues to support the conventional agrifood system.

Similar kinds of relationships are being developed within regional and community-based institutions. In our study of California alternative agrifood institutions, for instance, a number of organization leaders felt that significant progress was being made in working with local governments. One respondent said that the city in which they were working had, "turned its attitude 180 degrees from being against community gardens to now actively finding land for us to use, and encouraging us to apply for grant money that is out there to get gardens started." Another remarked that "the city officials changed the designation of our zoning from a soccer field to a sustainable agriculture education park, demonstrating their support of this project." Three-quarters of the organizations in the study reported working with government institutions. Government institutions at the city, county, and state levels are supporting the efforts of alternative agrifood movements. This combination of forces will help progress toward the long-term integration of the priorities of sustainability and sustenance.

Sustainable agriculture and community food security have become legitimate fields of discourse about problems requiring solution, and they have been the basis for developing policies and programs. They have catalyzed discussion and action on sustainability and community food security issues where little or no discourse on these topics had previously existed. Although their orientations in some ways oppose those of the dominant food and agriculture paradigm, these movements are making institutional inroads and expanding perceptions and practices of sustainable agriculture and community food security.

discourses, epistemologies, and practices of sustainability and sustenence

Alternative agrifood movements have made significant advances in integrating their priorities into traditional agrifood institutions and in developing new, alternative agrifood institutions. Sustainable agriculture and community food security discourses and practices have become a basis for a number of national and state programs and projects. New institutions like community-supported agriculture and farm-to-school programs are developing rapidly. These programs and institutions are recognizable and measurable forms of change in the agrifood system. This is not to say that the effort is complete or that it is always effective. While in some ways the alternative agrifood movements challenge and reconfigure perspectives and processes in the conventional agrifood system, there are other areas in which they may be reproducing them. To develop a more complete understanding of the role alternative agrifood formations are playing in shifting and reshaping the discourses, epistemologies, and practices of the agrifood system, we need to look behind the forms of the changes and to their content.

This content can be thought of as the "knowledge" produced by alternative agrifood movements and institutions. Eyerman and Jamison (1991) define knowledge in the context of social movements as the basic assumptions, ideas about the world, and topics and issues addressed by the movements. For them, the dynamic and mediating role that social movements play in the social shaping of knowledge is of central importance. The "cognitive praxis" of social movements creates new perspectives in science, ideology, and everyday knowledge.

Activists and scholars have attributed progressive characteristics and transformative potential to alternative agrifood movements and activities. Compared to the conventional food system, the constellation of alternative agrifood efforts is seen as creating an agrifood system that is more equitable, environmentally sound, and better for human health. For example, Clancy (1997) sees alternative food systems as pathways to adequate producer livelihoods, farmland and environmental resource preservation, equitable food-related employment, citizen engagement in food policy, food safety, and greater equity within the food system. According to Kloppenburg and others (1996), a sustainable food system is one that is sustainable, composed of people knowledgeable about the food system, locally based, as economically lucrative for farmers and farmworkers as off-farm labor, participatory, relational, just and ethical, regulated, sacred, healthful, diverse, culturally nourishing, seasonal, and more concerned with sustainability and equity than with profit.

How do the understandings being produced by alternative agrifood efforts support these kinds of claims? Raymond Williams's (1980) concept of dominant, residual, and emergent cultures provides a framework for pursuing this question. Dominant forms consist of those ideas, values, meanings, and strategies that have been produced and are inscribed in our current social system. They reflect the "common sense" of the system. Residual beliefs and practices are derived from an earlier stage of society and may reflect a different social formation than the present. An example of a residual belief is the Jeffersonian notion of the ideal farmer-citizen who was considered the paragon of virtue and foundation of democracy. Emergent ideas and practices are new social ideas, often produced by social movements. The emergent values, meanings, and strategies represent new or significant modifications of dominant forms.

All dominant ideas and practices were once emergent, but not all emergent ideas and practices become dominant. In the case of alternative food and agriculture, emergent forms are concepts and strategies that differ from and challenge dominant paradigms of the conventional agrifood system. They entail reformulations or transformations of the discourses, epistemologies, and practices of traditional institutions. At present, alternative agrifood movements and institutions contain all three forms. That is, while in many ways they differ significantly from conventional agrifood discourses and practices, in other ways they differ only slightly or not at all.

In this chapter I focus on dominant and emergent forms in the discourses, epistemologies, and practices of alternative agrifood efforts. In each area I first highlight emergent forms and then discuss the retention of dominant forms. In the section on discourse I discuss the inclusion of social issues and perspectives of social justice in the alternative agrifood movement. The section on epistemology looks at how alternative agrifood methodologies differ from conventional ones, focusing in particular on multidisciplinary and whole systems. The practices section brings up the issue of the neoliberal component in alternative agrifood strategies.

Discourses of Alternative Food and Agriculture

Discourse is "a specific ensemble of ideas, concepts, and categorizations that are produced, reproduced, and transformed in a particular set of practices and through which meaning is given to physical and social realities" (Hajer 1995: 44). The importance of discourse is that it recursively shapes how people's experiences are perceived and given meaning. "Discourses do not just reflect or represent social entities and relations, they construct or 'constitute' them" (Fairclough 1992: 3). The power of discourse is that it allows us to see in certain ways and not in others. Discourse determines what can legitimately be spoken about and who has the authority to speak and to whom. It also determines where and when a topic can be addressed. The insidious aspect of discourse is the extent to which it can operate outside our awareness. This opacity makes it a particularly effective shaper of reality. It is through discourse that dominant perspectives within organizations, institutions, and society in general are produced, reproduced, contested, and transformed (Fairclough 1994). In this chapter I use the term "discourse" to refer specifically to how sustainable agriculture and community food security are framed and defined.

Emergent Discourses

Now that sustainable agriculture and community food security have become established as legitimate discourses and integrated into conventional agrifood institutions, how are these concepts being articulated? To what extent and in what ways are social issues being included in alternative agrifood discourse?

In general, agricultural sustainability is defined quite broadly by both nongovernmental and governmental organizations. While this was a struggle as recently as ten years ago, today most groups include the three "Es" of environment, economics, and equity in their definitions and visions for sustainable agrifood systems. For example, a California-based consortium of public and private grant makers, the Funders Agricultural Working Group (2001) defines a sustainable food and agriculture system as one that:

- Protects the environment, human health, and the welfare of farm animals
- Supports all parts of an economically viable agriculture sector, and provides just conditions and fair compensations for farmers and workers
- Provides all people with locally produced, affordable, and healthy food
- Contributes to the vitality of rural and urban communities and the links between them

In our study of California alternative agriculture institutions, half the organizations we surveyed had a position on social justice. There is good reason to believe that many more have social justice as an implicit priority. For example, there were respondents who said their organization had no position on social justice, yet the primary work of the organization involved improving conditions for low-income people, activities clearly within the domain of social justice work. In still other cases, some organization leaders who stated that their organization did not have a position on social justice, and that their organization did not work on social justice issues, spoke elsewhere in the interview about the need to stop putting profits before feeding hungry people or the injustice of the commodification of food.

This shift in discursive orientation toward social issues is mirrored in sustainable agriculture programs (discussed here because of their relatively long histories). The discursive content of the mission and vision statements of sustainable agriculture programs has become broader and more complex, as illustrated by the inclusion of social factors, which were previously overlooked in sustainable agriculture programs. Initially, sustainable agriculture was defined in terms of environmental protection and farm-level profitability. For example, in its first brochure describing its low-input sustainable agriculture (LISA) program, the USDA (1988) stated that "low-input, sustainable agriculture addresses multiple objectives—from increasing profits to maintaining the environment—and may incorporate or build on multiple systems and practices such as integrated pest management and crop rotations." One of the first evaluations of the USDA's activities in this program included categories for environmental and economic impacts, but not social impacts (House 1988). Then, in 1990, the name of the program was changed from "Low-Input Sustainable Agriculture" to "Sustainable Agriculture Research and Education" (SARE).This name change recognized that sustainable agriculture is "broader in scope than simply reducing pesticides and fertilizers but includes the need to enhance the economic viability of farm operations and the quality of life for farmers and society as a whole" (GAO 1992b: 2). This vision is notable for its inclusion of social factors and its attention to agricultural issues beyond the farm gate. While SARE continues to prioritize improving profitability for farmers and protecting natural resources, these goals are now supplemented by two others: (1) improving the quality of life for farmers, rural communities, and society; and (2) enhancing the quality of life for farmworkers, agrarian industry workers, and consumers (SARE 1998b). This shift represents a significant discursive reorientation from exclusively technical production to the inclusion of social priorities.

Similarly, the University of California Sustainable Agriculture Research and Education Program (UC SAREP) has expanded its mission since its early years, moving from a productivist to a broader food-and-agricultural-systems orientation. In its first progress report, UC SAREP director Bill Liebhardt wrote that UC SAREP's goal was to "provide the agricultural community with new, science-based options that ensure both economic viability as well as environmentally responsible production" (UC SAREP 1990: 1). The report made no mention of social issues or priorities. But by 1991 UC SAREP had developed a comprehensive definition of sustainable agriculture that integrated "three main goals—environmental health, economic profitability, and social and economic equity" (UC SAREP 1991: 1). It identified agricultural problems as including social categories such as the living and working conditions of farmworkers and the social and economic conditions of rural communities. In its 1997 long-range planning process, UC SAREP developed a "deliberately" broad mission statement: "UC SAREP provides leadership and support for scientific research and education to encourage farmers, farmworkers, and consumers in California to produce, distribute, process, and consume food and fiber in a manner that is economically viable, sustains natural resources and biodiversity, and enhances the quality of life in the state's diverse communities for present and future generations" (Liebhardt 1997).

The W. K. Kellogg Foundation's Integrated Farming Systems (IFS) initiative reflects a similar progression. The IFS initiative was designed to solve problems such as environmental degradation, decreased profitability for farmers, food safety issues, the demise of rural communities, and a lack of opportunities for new farmers (Hesterman and Thorburn 1994: 134). Accordingly, Phase 1 projects focused on the development and implementation of sustainable agricultural practices. Projects funded under Phase 2 of the initiative, however, were significantly more comprehensive in subjects addressed, strategies outlined, and participants involved than earlier projects (see Table 5). The Phase 2 projects included topics such as exploring long-term implications of sustainability, market changes, the impact of public policy, and changing consumer behavior. A key discursive change is inclusion of the word "food" in the name of the initiative— Integrated Food and Farming Systems (IFFS)—reflecting a new emphasis on food security as part of an integrated approach to agrifood systems and policies.

For all three of these programs, there has been a clear progression from a focus on the environment to the range of economic and social issues inherent in the agrifood system. Sustainable agriculture and community food security programs have made significant progress in increasing consciousness about and changing the terms of discourse in sustainable agriculture and food security in challenging institutional environments.

Retention of Dominant Discourses

While the discourse of sustainable agriculture discourse differs from that of conventional agriculture—and increasingly so—there are discursive orientations that remain closely aligned with those of conventional agriculture. Examples include an emphasis on farmers (discussed in Chapter 6), and the circumscribed ways in which social issues are conceptualized and framed. Early on, social issues were rarely considered in definitions of sustainable agriculture, as illustrated in early definitions by the SARE program. When social factors were mentioned, they were often described and framed in safely vague terms of what is "socially responsible" or "socially acceptable" in reference to environmentally and economically sustainable institutions and practices. For example, Crosson (1991: 14) wrote that "a sustainable agricultural system is one that can indefinitely meet demands for food and fiber at socially acceptable economic and environmental costs." This begs the question, "socially acceptable for whom?" Where a "who" was indicated, it generally referred to an abstraction of future generations rather than to those living in the present time.

In reviewing definitions of agricultural sustainability, Dicks (1992: 191) concludes that "the common thread of all the definitions is the requirement that future generations have access to the same or better quality and quan-

Table 5 Kellogg Integrated Food and Farming Systems (Phase 2) Projects

Organization	Purpose
Alternative Energy Resources Organization, Helena, MT	Create community-based food and farming systems that foster the social, environmental, and economic health of local communities and agriculture.
Center for Sustainable Systems, Berea, KY	Build capacity of leaders for systems-level change toward a sustainable food and farming system.
Henry A. Wallace Institute for Alternative Agriculture, Greenbelt, MD	Design a comprehensive future national policy framework for food, agriculture, and rural sustainability.
Institute for Policy Studies, Washington, DC	Improve leadership capacity in nonprofit, public interest groups committed to sustainable agriculture through a Social Action and Leadership School.
Keystone Center, Washington, DC	Create a new dialogue project regarding the future of agriculture in terms of current trends toward industrialization, vertical integration, and contract farming.
Michael Fields Agricultural Institute, Madison, WI	Provide media training to IFS grantees and those involved with the Campaign for Sustainable Agriculture and the Sustainable Agriculture Working Groups.
Minnesota Project, Inc., Pine Bush, NY	Bring together leaders within the Sustainable Agriculture Working Groups and the Integrated Farming Systems Network with the National Campaign for Sustainable Agriculture to identify common objectives and design an action plan.
Ohio State University, Marysville, OH	Develop the leadership and initialize the activities of the IFFS Network through leadership and organizational development of the IFFS Steering Committee

Source: W. K. Kellogg Foundation n.d.-a.

tity of food, fiber and environmental amenities as we have today." This theme of maintaining the system and its distributional effects in its current form goes back to the beginning of discussions about agricultural sustainability. One of the first to articulate ideas about sustainable agriculture, Douglass (1984: 25) writes: "Agriculture will be found to be sustainable when ways are discovered to meet future demands for foodstuffs without imposing on society real increases in social costs of production and without causing the distribution of opportunities or incomes to worsen. . . . The goal is to have an agricultural system the results of which are no worse than the existing system's." Talking about the future rather than the present deflects attention from the need to discuss contentious issues such as human rights and access to resources. Yet, as Lipietz (1995: 149) points out, the rights of future generations can only be achieved if we work to ensure those rights within present society.

The W. K. Kellogg Foundation is notable in that it conceptualized sustainable agriculture as including social considerations from the beginning. However, these social considerations are somewhat limited. In a paper written to describe the Foundation's vision of agricultural sustainability, the two goals of its Integrated Farming System Initiative were to:

> 1. Help farmers find and adopt integrated and resource-efficient crop and livestock systems that maintain productivity, that are profitable, and that protect the environment and the personal health of farmers and their families.

> 2. Assist people and their communities in overcoming the barriers to the adoption of more sustainable agricultural systems so that these systems can serve as a foundation upon which rural American communities will be revitalized. (Hesterman and Thorburn 1994: 133)

Nongovernmental organizations may have an easier time embracing and using discourses of social equity and social justice, and many do. However, even people and organizations that do use the term "social justice" can have widely different definitions for the term. Early works by Gips (1988) and Freudenberger (1986) that include notions of justice in their conceptions of sustainable agriculture illustrate this variation. While Gips includes issues such as workers rights and humane treatment of animals when using the term "justice," Freudenberger uses it to mean providing opportunities and support structures for individual farmers. We found similar differences in the meaning of the term "justice" during our study of California alternative agriculture institutions (Allen et al. 2003). We asked leaders of alternative agrifood institutions across the state of California, "In your personal view, what would a socially just food system look like?" There were five basic patterns that emerged in analyzing people's responses. They defined "social justice" in terms of economic equity, as relational or proximate, in terms that centered on the farmer, in relation to health or the environment, or in terms of accessibility to food. Respondents were evenly distributed among these categories.

Twenty-two percent of the respondents described a socially just food system as one that is economically equitable. Criteria they used included the fair compensation of labor, common ownership of land, and a food system in which everyone's basic needs were met regardless of ability to pay. Some organization leaders pointed out that fundamental social relations of economic privilege needed to change in the food system in order for it to become socially just. Beyond this, several respondents emphasized that developing a socially just food system would only be possible with larger changes in that direction throughout the whole of society, that is, not just within the food system. One interviewee said, "The food system can only be as socially just as the bigger economic, social system within which it exists." Related to economic equity was the category of food accessibility, cited 19 percent of the time. Those who mentioned characteristics in this category were concerned that food be available primarily to low-income people. The category of food accessibility was separated from that of economic equity because it presupposes a society in which there would be disparities in income. While those who defined social justice in terms of food accessibility wanted to make sure that no one in society went hungry, those concerned with economic equity were focused on more basic changes through which the category of low-income people would cease to exist.

More than half of the organization leaders defined social justice in ways that do not relate to equity or accessibility. Nineteen percent of responses defined a socially just food system as local or proximate, that is to say, one in which most of people's food would be sourced from regional farms. This definition meant more to the respondents than spatial proximity, however. These respondents emphasized the importance of farmers and consumers having close, personal relationships and consumers "knowing where their food comes from." Another 21 percent of the responses focused less on the location of farms or the relationship between farmers and consumers, but defined a socially just food system as one in which small or family farms and businesses would be viable. In this framing of social justice, respondents included goals such as increasing the share of the food dollar retained by farmers and making food more expensive for consumers. There was a strong sense that farmers should not be price takers or subject to the whims of consumers. One respondent said, "The food system should be owned by the farmer, and run by the farmer." In another 19 percent of responses, a socially just food system would be one in which food was grown in a way that did not degrade the environment and which provided an opportunity for people to buy healthy, nutritious food if they so chose.

In other ways, the discursive category of justice has been invisible or studiously avoided. For example, Charles Benbrook, a prominent leader in the sustainable agriculture movement,[1] stated that "the fundamental social responsibility of organic agriculture is improving the health of the soil—there is universal consensus that farmers and agriculture systems have to take care of the soil. But there is no consensus on the nature of justice and what equity is and how the state should intervene in the structure of agriculture" (quoted in Gershuny and Forster 1992: 8). Of the half of the California alternative agrifood organizations that had no position on social justice, many said that they simply didn't have one, or that they did not think in terms of social justice and injustice. However, some responses carried a tone of exasperation with the subject. Responses included comments such as "We don't have time to get into that," or "No, we're here for farmers." Some stated that the issue of social justices was too contentious. One response reflected a kind of disdain for low-income food programs in answer to this question, stating that what is unjust is people spending food stamp money on soda pop and potato chips.

Although this is beginning to change, American alternative agrifood movements have also tended to be somewhat silent about the structure of social inequities. For example, they prioritize fair returns to farmers but have little to say about equitable conditions for agricultural labor. Similarly, they decry the loss of family farms, but they disregard the gendered and racialized structure of family farming in U.S. agriculture. The movements mostly overlook basic questions about the fairness of the current agrifood system, submerging problems such as class, race, and gender asymmetries. An example of a shift in this regard is a front-page article on the first issue of the IFFS newsletter addressing efforts to stem the loss of black farmers (Integrated Food and Farming Systems Network 1998). This is one of the few times this issue has received attention in discussions within the sustainable agriculture movement about the loss of small family farms.

Without adequate analysis, these central categories of injustice upon which the present agrifood system is built will ultimately be reproduced in sustainable agriculture and community food security or embodied in dubious surrogate discourses such as communities and family farms. While the kinds of discursive changes being made through sustainable agriculture and

---

1. From 1984 to 1990 Charles Benbrook was executive director of the Board on Agriculture of the National Academy of Sciences, which published the major report, *Alternative Agriculture*, in 1989.

community food security programs represent a crucial stage in the development of these movements, they need to go further in addressing the issues of social justice and social inequity. This is essential to avoid the watered-down, institutional conceptions of sustainability and food security and to address the fundamental issues upon which agricultural sustainability and food security depend.

The disciplinary force of discursive practice comes from the assumption that others will utilize the same discursive frame (Hajer 1995). Alternative agrifood activists can believe that their concerns are being addressed within that frame, and may even come to think that what is proposed within that frame is all that reasonable people can do. Institutional conversation involves constraints on what counts as legitimate contributions to goals and projects. This discursive approach then gets inserted back into activist, academic, and public discourse in ways that prescribe the boundaries of problem constructions and solutions and even what one dares to say at a meeting. Although these kinds of discursive practices are arbitrary, they are also opaque—people may not recognize these ways of seeing a problem as temporary moments of political positioning. Instead, they may assume that this is simply the correct way to talk about a particular subject.

This type of discursive construction is both a cause of and product of institutional success. Constraining discourses and strategies are sometimes internal to the movements themselves. Limiting discourse to less controversial topics increases acceptance. At the same time, the process of institutionalization and the resulting programs may shape and reshape the meanings and practices of sustainable agriculture and community food security in ways that may actually run counter to the goals of long-term sustainability and social equity. The discursive content of social movements is, of course, a dynamic rather than static process. New issues emerge, tensions come up, new members join, and new opportunities arise. All of these can play a part in reshaping the discursive basis of the movement.

Epistemologies of Alternative Food and Agriculture

A similar tendency to challenge conventional frames—but within relatively narrow boundaries—often characterizes research efforts within alternative agrifood programs. Epistemological approaches, categories, and concerns are absolutely central to shaping the food and agriculture system because they determine which problems are important and how they are defined and addressed. The power of epistemology—the process through which people come to know their world—is that it can limit the ways in which solutions are derived, which options are considered available and appropriate, and what types of changes are likely to take place. As O'Neill (1986: 91) observes, "In adopting certain categories for social inquiry we also adopt a certain view of the social world, of its problem areas and of its fixed points, of the actions it makes available and ways in which their results are constrained." The problems deemed important and the way in which questions are posed in sustainable agriculture and community food security prescribe certain possibilities for change and proscribe others.

There are basically two types of research—one dedicated to improving performance (e.g., increasing production) and the other dedicated to improving the human condition (Fairclough 2001). In the current research funding system, the first of these has much higher priority. For example, from 1973 to 1996, the small amount of funding going toward the social sciences in the United States dropped by 40 percent, from 8.0 percent in 1973 to 4.8 percent in 1996 of total federal/nonfederal funding sources (Rapoport 1998). This trend is particularly clear in agrifood system research. For example, in 1987 only one percent of USDA research funds was spent on projects in sociology or anthropology (National Science Foundation 1989). The USDA recognizes that the problems facing rural America are largely due to social, economic, and cultural conditions and, as such, "cannot successfully be addressed solely with the knowledge generated by the biological or agricultural sciences" (USDA 1993a). However, this has not translated into funding for social issues research. Support for social science research in agriculture declined further between the 1980s and the 1990s (USDA 1993a). While support for "improving the human condition" has not been a hallmark of alternative agrifood programs, they nonetheless involve research programs that engage expanded problem statements and epistemologies as compared to traditional agricultural research.

Emergent Epistemological Approaches

Programs in sustainable agriculture and community food security often reflect expanded epistemological frameworks of how agrifood issues are considered and studied. As compared to traditional research programs, they place greater emphasis on environmental concerns, multiple disciplines, and whole-systems approaches.

Traditionally, American agrifood research institutions have focused on harnessing nature to produce food. Since the eighteenth century, the dominant epistemological model (inherited from Francis Bacon) has stressed that, through science and technology, humans can dominate and master nature by advancing knowledge. Following this Baconian imperative, publicly supported agricultural science has attempted to subdue and conquer nature to serve human interests through mechanistic, reductionist, and fragmented approaches to understanding and shaping the natural world. Sustainable agriculture scientists privilege "nature" more than their conventional agriculture counterparts and generally reject the Baconian view of most of the agricultural sciences that nature is an object solely to be controlled and manipulated to serve human needs.

Although they are usually located in traditional agricultural science departments, sustainable agriculture researchers have broadened the agricultural science agenda considerably. The major change has been that increased attention is given to maintaining natural resource conditions of production in agriculture. This has two sides. One is to minimize the degradation of the environment by reducing the need for chemicals, thereby reducing groundwater pollution and protecting wildlife (as well as reducing input costs). The other is to conserve resources upon which agriculture depends, such as soil, groundwater, and fossil fuel energy (by reducing dependence on chemicals and other energy-based inputs). These priorities are reflected in the research and education projects funded by the SARE and the UC SAREP programs.

One criticism of conventional agricultural science is that it takes single components of agricultural systems and studies them in isolation. The shortcoming of this approach, some have said, is that it is blind to how alterations in one component might affect other components. In traditional agricultural research and education there is a long-standing separation of the component parts of the agrifood system. Different departments—for example, soil science, plant pathology, animal science, and agricultural economics—each focus on their "part." This dissection of the agrifood system into disciplinary departments is reflected in the narrow focus of conventional agricultural projects and approaches.

Research in sustainable agriculture, on the other hand, is multidisciplinary; that is to say, it employs a number of different disciplines brought together under a single project or agenda. In arguing that multidisciplinary research is integral to sustainable agriculture, Lockeretz (1991) points out that multidisciplinary research promotes attention to topics such as the environmental and social consequences of agricultural systems that are generally overlooked in conventional agricultural research. Stevenson and others (1994) see multidisciplinary research as addressing three different needs:

(1) a need for systems-oriented research, (2) a need to include farmers and nonuniversity organizations in the research circle, and (3) a need to communicate the results of research beyond the immediate agricultural community to consumers and policymakers.

Interdisciplinary approaches can be thought of as somewhere between the relatively loose connection between the disciplines characteristic of multidisciplinary approaches and the narrow focus of conventional agricultural research. While not strictly research oriented, community food security efforts are interdisciplinary in this way. Community food security is rooted disciplinarily in urban planning and community development rather than in agricultural or nutritional sciences. It is at its core an interdisciplinary, integrated approach and represents a significant departure from the predominant highly scientized and statistical approaches to hunger, nutrition, and toxicology that are familiar from hunger-relief and related development projects.

An approach that integrates food access into urban planning is starting to attract the interest of major research institutions (although not traditional agricultural departments). For example, the Department of Urban and Regional Planning at the University of Wisconsin, Madison, teaches a rare graduate course on food-system planning. Food security is addressed as part of student projects. In 1997 the class produced a report that examined why planners had paid so little attention to the food system, why the food system requires the attention of planners, and how planners can play constructive roles in food-system development (Pothukuchi and Kaufman 2000).

Community food security is also based on a whole-systems approach to agrifood issues. This approach critiques traditional ways of looking at food security issues as fragmented and lacking in overarching vision and coherence. The whole-systems approach asserts that a lack of coherent vision poses a major obstacle to the development of long-term food security and sustainable food systems (Fisher and Gottlieb 1995). Community food security uses a food-systems approach to comprehensively identify problems and articulated solutions, not a traditional antihunger orientation. For example, Canada's Toronto Food Policy Council frames food security as a health issue in which hunger and poverty are viewed as part of the larger issue of public health, a perspective that makes access to food not only equitable but economical as well.

This systems approach is also present in sustainable agriculture research programs, both in an emphasis on farming systems and on "whole systems." By emphasizing systems research and environmental impacts, the USDA

SARE program's strategies have moved substantially beyond those employed in traditional agricultural research. Sustainable agriculture researchers focus more on integrated farming systems in their efforts to mitigate unintended environmental consequences of agricultural practices. Much of the early research in sustainable (low-input) agriculture operated under a straightforward "input substitution" paradigm. That is, it replaced "artificial" chemical inputs with "natural" environmentally benign alternatives, "without challenging either the monoculture structure or the dependence on off-farm inputs that characterize agricultural systems" (Rosset and Altieri 1997: 283–84). Many sustainable agriculture researchers now argue for an approach based on the science of agroecology, which takes the whole system as its unit of analysis. It is a "merger of analytical methods, concepts, and data from ecology with more traditional approaches for studying agricultural problems" (Lockeretz and Anderson 1993: 70). Agroecological approaches, which include principles of diversity, adaptability, durability, and symbiosis, are becoming more prevalent in agricultural research. A recent article by leaders in the fields of sustainable agriculture and agroecology points out that it is no longer sufficient to study agrifood systems using standard ecological, agronomic, and economic values and approaches (Francis et al. 2003). They redefine agroecology as the integrative study of the entire food system, including ecological, economic, and social dimensions

Initially, very few proposals that employed "integrated systems" approaches were submitted to SARE, partly because those submitting the proposals generally lacked experience with systems research, because component research provides better rewards in research institutions, and because systems research is more difficult to organize and carry out. Interdisciplinary researchers are further restricted by disciplinary constraints internal to intellectual work (Haila and Levins 1992). SARE director Jill Auburn (2002) reports that one of the limitations to doing holistic research is that it is hard to design a holistic methodology that also satisfies people from multiple disciplines. SARE regional programs responded by increasing systems-type research, limiting proposals to those that included integrated systems, holding a grant-writing workshop on how to develop systems projects, and developing guidelines for evaluating whole-systems proposals. SARE began to emphasize the trinity of the environmental, economic, and social aspects of food and agriculture and articulated a systems approach to sustainability: "Agriculture is often viewed as consisting of three types of systems: economic, ecological, and social. Sustainable improvement in agriculture—usually thought of in terms of farm profitability, environmental stewardship and quality of life for farm families and rural communities—must be based on these interlocking aspects of agriculture" (USDA 1997a).

A number of current trends that incorporate farmers into sustainable agricultural research (e.g., "participatory research," "on-farm research," "farming systems research") were inspired by the tradition of participatory action research. Participatory action research has been practiced most commonly in impoverished regions, where it has been part of a struggle for the emancipation of poor and powerless people. As such, its aim is not simply to gain technical knowledge (although that is part of the project), but also to empower people to change their situations. Participatory research is used for social research as well as for on-farm research. Research on social systems, such as marketing, political organizing, and knowledge exchange, is an important part of participatory research because it is through these forms of organization that people empower themselves politically. There has been an increasing number of efforts to apply the philosophy and methods of participatory action research to the study of farming production methods, such as in California's biological farming systems research projects like BIFS (Biologically Integrated Farming Systems) and BIOS (Biologically Integrated Orchard Systems). These types of research projects tend to be locally specific and tailored to farmers' needs.

The SARE program has also gone to great lengths to meet the real rather than imputed needs of farmers. A major criticism of the traditional agricultural research system is that information has flowed from universities to the farms without paying sufficient attention to farmer problems and knowledge (Hassanein 1999). SARE counteracts this approach to agricultural research and education by involving producers in setting priorities and funding decision making (SARE 1995). The priority given to farmer involvement in research is evident from the requirement that farmers be on the regional administrative councils, the creation of a producer grants program to fund on-farm research and demonstration projects, and the establishment of demonstration projects.

When sustainable agriculture was first becoming institutionalized in the late 1980s, Buttel and Gillespie (1988) predicted that the sustainable agriculture research programs located in dominant agricultural institutions would be merely "born-again agronomy," that is, applied commodity research conducted by scientists in traditional agricultural fields such as soil science and entomology. In the beginning, this appeared to be the case. For example, UC SAREP began with a fairly narrow focus, with funding directed almost exclusively at production or marketing. In UC SAREP's first request for proposals, four priorities areas were designated: "evaluation of organic and other non-synthetic chemical farming practices"; "comparison of food produced with and without synthetically compounded chemicals to examine nutritional quality, toxicity, and shelf life"; "the process of conversion to reduced use of chemicals in large, medium, and small scale farming"; and "development of a model to evaluate long term environmental and economic costs of chemical contamination, toxic wastes, and chemical residues in agriculture" (UC SAREP 1986: 2). Three years later, priorities were little changed. UC SAREP's 1989 request for proposals invited projects in the areas of weed management and control, alternatives to fungicides and fumigants, soil nutrient alternatives, and breeding research. The proportion of funding for research in these areas has, however, decreased over time, allowing space for projects in community development and public policy.

Although programs are still overwhelmingly focused on traditional agricultural production topics, sustainable agriculture has broadened this agenda to include issues beyond the farm gate. In response to a request by the program's public advisory committee, in 1990 UC SAREP added an emphasis in economic, social, and public policy research and hired a specialist to help the program better address issues of economics and public policy (UC SAREP 1993). The social science projects funded by UC SAREP are quite broad in scope, including those focused on land tenure arrangements, farmworker development, community food systems, and public policies. For SARE, the relative proportion of funding allocated to community development as compared to agricultural production has increased since the program's inception. Thus, the programs' expanded programmatic statements about sustainable agriculture and their actual funding priorities have become more closely aligned over time.

CASA also focused on multiple aspects of the food and agriculture system. For example, one of the project objectives is to identify barriers to sustainable food and agricultural systems. This includes not only the usual barriers to the adoption of sustainable farming practices, but "the full range of barriers to sustainability: technical, social, economic, institutional, policy and cultural." In pursuing its objectives with regard to educational strategies, CASA includes not only projects focused on farm practices but also projects on topics that span the whole food system, including the integrated social and environmental aspects of sustainability and food and environmental movements.

Retention of Dominant Epistemological Approaches

The United States has the largest public agricultural research system in the world, dating back to the latter part of the nineteenth century, when Congress established the U.S. Department of Agriculture (USDA) and agricultural colleges in every state.[2] Although sociologists and other social scientists have played a significant role in the emergence, institutionalization, and design of sustainable agriculture (Buttel 1993b), the primary epistemological terrain and actual practice of sustainability have been largely confined to the natural sciences. As Buttel observes, the "impetus for and the legitimacy of sustainability come from natural science, natural scientists, and natural science data" (Buttel 1993a: 26). Sustainable agriculture research and education programs are placed in conventional agricultural departments at traditional agricultural universities and are staffed by scientists with expertise in traditional agricultural fields, which are isolated from the social sciences.

This is reflected in public sustainable agriculture programs. In both SARE and UC SAREP the overwhelming majority of research dollars for sustainable agriculture goes to natural sciences and focuses on farm-level projects and production innovations, while a much smaller amount is devoted to social constraints and possibilities. This funding imbalance in sustainable agriculture reproduces the traditional asymmetry between social and natural science in food and agriculture as a whole. Many sustainable agriculture research and education projects focus largely on environmentally friendly versions of what agricultural science already does—improve production practices on farms. Research on nonagronomic topics in sustainable agriculture is often limited to topics such as evaluating how farmers' values and attitudes encourage or block adoption of sustainable technologies and developing new markets for farm products. UC SAREP projects focused on social issues were in the minority, both in number and share of

---

2. In 1862 Congress passed the Morrill Land-Grant College Act, which provided 30,000 acres of federal land to each state for each representative and senator. The land was to be sold and the revenue used to establish colleges that would offer courses in agriculture and the mechanical arts. From the beginning, the idea was to increase the role of agriculture in the national economy through scientific discovery and the adoption of new technologies. In 1887 the Hatch Act provided annual funds to each state to support experiment stations to conduct scientific research in agriculture. In 1914 Congress passed the Smith-Lever Act, which provided states with matching funds to develop a system to extend the results of scientific research to rural people.

funding, when placed alongside projects on, for example, marketing, soil fertility, or pest management, which are directed at improving farm-level production and profits. Between 1987 and 2001, 85 percent of SAREP's grant funding was allocated to agricultural production projects, and 6 percent went to projects focused on community development and community food systems.

SARE's project choices have also lagged behind changes in the program's broad views on the problems and definitions of sustainability. While the SARE program incrementally expands its definition of sustainable agriculture to include social as well as production-oriented priorities, social priorities get defined in fairly narrow terms. Projects funded in SARE's social category focus primarily on subjects such as local food production and marketing, farmer quality-of-life comparisons, and the implementation and adoption of sustainable agricultural practices. Even these receive a small fraction of overall project funding; 76 percent of the projects SARE has funded have been in standard agricultural production areas such as pest management, animal production, crop production, horticulture, and integrated farm/ranch systems. The remainder of the funding is divided among areas such as economics and marketing, natural resource protection, and community development. SARE funding records show that agricultural production projects were funded in every year, but "quality of life" projects were funded in only four of the ten years. Even the community development and quality of life projects focused primarily on farmers and agricultural communities, leaving out workers in other sectors of the agrifood industry, both rural and urban.

The proportion of funding allocated to these various areas deviates from the sustainability priorities articulated by SARE. Nonetheless, SARE considers them to comprise a wide scope of the types of projects needed to achieve sustainable agriculture. For example, in the beginning of its 1998 ten-year report, a chart intended to "demonstrate the diversity of projects SARE has undertaken to advance sustainable research and education" (SARE 1998c: 5) was introduced. Yet virtually all of the projects displayed focused on agronomic techniques and practices, reflecting a disconnect between discourse and practice. SARE director Jill Auburn reports that it is a struggle to maintain emphasis on the social "third leg of the stool" and to talk about social issues in a way that resonates with rather than alienates farmers (Auburn 2002).

This lack of attention paid to the social aspects of sustainability is due not only to funding priorities and disciplinary divisions but also to sustainable agriculture's acceptance of conventional agriculture's epistemological assumptions. Sustainable agriculture is heir to the epistemological biases of conventional agriculture. Sustainability problems have been addressed through the traditional Western epistemological tools of positivist natural science and neoclassical economics, with standard agronomic and economic categories used as measures of success. At the same time that it is recognized that environmental problems have resulted in part from conventional agricultural practices, there is little critique of the epistemological approaches underlying agricultural science. In these domains, epistemology is rarely considered, much less contested. Even sustainable agriculture policy analysts tend to reify science itself as a somehow autonomous body of knowledge. For example, Youngberg and others (1993) write that advancing the sustainable agriculture movement requires replacing the "symbolism" of sustainable agriculture with "scientific facts."

The privileging of natural science in sustainable agriculture research is exacerbated by a general lack of interchange between social scientists and natural scientists in the university (Busch and Lacy 1983). Limited attention is paid to integrating underrepresented disciplines such as sociology or anthropology into sustainable agriculture programs. Publicly funded agricultural research institutions have been isolated and insular, and funding for agrifood research is largely confined to agricultural departments of land-grant universities and the USDA's Agricultural Research Service (Goodman 1997). One important issue in multidisciplinary research is the degree to which the different disciplines are integrated. A team consisting of an agronomist, an entomologist, and an ecologist would come up with very different explanations of a situation than a team consisting of an economist, a sociologist, and a political scientist.

However, Western science is not a universally considered and valued epistemology, and working toward sustainability requires using many types of cognitive maps (Redclift 1993). This is not to say that Western science is wrong, but that its perspective is partial and needs to be supplemented with the equally partial insights of other ways of knowing (Kloppenburg 1991). Recent trends in science studies and the philosophy of science have emphasized the partial and privileged perspective of the scientist's view, which is necessarily limited by what philosophers call "objectivism." This approach presupposes not only an epistemological break between subject and object, but also a social separation between observer and observed. In other words, it overlooks the social conditions that make science possible.

Much research in sustainable agricultural research is of the same technical, production-oriented nature as that in conventional agriculture. The premises of sustainable agriculture tend to mirror those of conventional agriculture in that they generally exclude humans and social institutions as analytical subjects. In general, working toward agricultural sustainability is still seen primarily as a natural/technical process of people interacting with nature, rather than as part of a complex web of social relations. This approach gives rise to an expectation that tinkering with conventional agricultural science can yield the answers to problems of sustainability in agriculture. In any agricultural system people interact first and always with *other people*, in addition to their interactions with nature. This absence of attention to the social causes of nonsustainability and food insecurity can severely limit the efficacy of proposed solutions. In natural-science approaches to solving agricultural problems, social relations are not generally considered relevant. "Science treats nature as external in the sense that scientific method and procedure dictates an absolute abstraction both from the social context of the events and objects under scrutiny and from the social context of the scientific activity itself" (Smith 1984: 4). Science, of course, is never "pure," in the sense that research is always embedded in social relations, including those that make science possible in the first place. Research problems are constituted socially through the selection of research priorities and allocation of financial resources to some areas and not others.

Sustainable agriculture research has been guided by an implicit assumption that social relations in food and agriculture are fixed and largely beyond human control. Social relations are treated as constants and interactions with nature as manipulable variables. Overlooking the centrality of human action has led sustainability advocates and researchers to define problems primarily in terms of nature and environment. Priorities in sustainable agriculture programs reflect general assumptions that environmental change takes precedence over social change, and that achieving agricultural sustainability is possible without changing social relations. For example, the National Research Council's (1989) influential report on alternative agriculture shows a lack of interest in factors other than technical production. In this report "alternative" refers to biological and technological alternatives to conventional agricultural practices, not to alternative social and economic arrangements. Generally, there has been little or no serious investigation into the social, political, and economic relations that are needed to encourage sustainable agriculture.

The fact that agricultural science is partial, value-laden, and socially constructed does not mean that it does not produce useful knowledge. Both the biological and physical agricultural sciences have yielded important agronomic and technical solutions to agricultural production problems. However, they are hardly equipped to deal with the fundamental causes of and corresponding solutions to the problems that sustainability poses for current agricultural systems. In his pioneering agroecosystems work, *Agroecology: The Scientific Basis of Alternative Agriculture*, Altieri found that, in developing sustainable agroecosystems, it is "impossible to separate the biological problems of practicing 'ecological' agriculture from the socioeconomic problems of inadequate credit, technology, education, political support, and access to public service" (Altieri 1987: 196). While research on biological pest control, rotational grazing, intercropping, cover crops, and nutrient management have necessarily provided farmers with less environmentally damaging production strategies, social science research is needed to create long-term solutions. A natural-science approach is applied even when social change is the goal. As noted by Hamlin (1991: 508–9) in an article on sustainable agriculture, "One cannot help but be struck by the degree to which proposed social changes are sanctioned in appeals to biology, toxicology and the earth sciences, rather than notions of justice, effective government or progress."

And while alternative agrifood efforts are definitely expanding the units of analysis, the focus and activities are still concentrated on technical aspects of farm-level production. In both the SARE and UC SAREP programs, research and education dollars are spent primarily on improving agricultural production methods. Like conventional agriculture, sustainable agriculture continues to overlook the equally important distribution, exchange, and consumption components of the agrifood system. This orientation became established early on in the movement, despite its emphasis on whole-systems approaches. More than twenty years ago the Cornucopia Project (1981) undertook a comprehensive study of the U.S. food system. It included many of the issues that span the sustainable agriculture and community food security movement. In addition to focusing on environmental problems in agriculture and market concentration, the published study *Empty Breadbasket?* had chapters on nutrition and health, urban food systems, and food assistance. Even with its broad focus and call for research on "sustainable food systems," the areas recommended for research were all agronomic, such as soil management, reduction of agrichemical use, and crop rotations.

Part of the problem is that what is meant by "whole system" in epistemological terms is often only the farming system. For example, a recent article on sustainable agriculture framed "total food-systems research" in terms of "new food marketing relationships" that would enable farmers to produce more value and retain that value for themselves and the local community (Kirshenmann 2002). Similarly, while Edwards (1990a) argues the importance of integrating component parts of sustainable agricultural systems, he limits these components to on-farm elements such as fertilizers, pesticides, crop rotations, and farm equipment. Edwards states that increasing knowledge of the main farm inputs and how these practices interact will form the basis of developing agricultural systems that increase profitability for the farmer and reduce environmental problems. This perspective also predominates in the National Academy of Sciences' approach to alternative agriculture, as seen in its 1989 report, *Alternative Agriculture* (NRC 1989).

In its strategies for achieving sustainability the W. K. Kellogg Foundation explicitly works to move beyond the usual technical approaches to address human and institutional factors. Foundation leaders point out that many agricultural problems "are actually human challenges that cannot be successfully addressed solely with technological solutions" (Hesterman and Thorburn 1994: 132). Only one of the four barriers to sustainability identified by the Foundation is technical—how to manage pests and provide adequate plant nutrients while reducing negative environmental impacts. The other three barriers to sustainability identified by the Foundation are (1) political and economic (e.g., federal farm programs), (2) institutional (e.g., disincentives for sustainable agriculture research and education, the lack of a definition of agricultural sustainability, and an excessive focus on technology rather than the process of change), and (3) personal (e.g., disbelief in the practicality or profitability of adopting sustainable practices). The second-phase projects dealt with topics such as the long-term implications of sustainability, market changes, the impact of public policy, and changing consumer behavior. The conceptual framework for solving sustainability problems was greatly expanded under the Foundation's initiative, which discusses social barriers to the adoption of sustainable agricultural systems. Nonetheless, its funding priorities continued to emphasize technology and agricultural businesses:

> Priority is being given to projects that (i) develop, test, and validate technologies that support the development of more resource-efficient, integrated farming systems (including validation of economic feasibility); (ii) include an innovative educational component to promote the

adoption of these technologies; (iii) explicitly address the challenge of enabling effective communication and responsible decision-making among stakeholders in our agricultural communities; and (iv) formulate a strategy to develop leadership capacity within farm families and businesses and then use that leadership to shape the dialogue between farmers and ranchers and the nonfarm members of the community. (Hesterman and Thorburn 1994: 133)

Even though the on-farm transformation of resources into food and fiber is a core process of the food and agriculture system, it is but one of many components. Food systems include not only production but also distribution, processing and preservation, preparation and consumption, and waste disposal (Dahlberg 1993). All of these parts of the food system operate within a larger policy, research, and economic framework that affects production and distribution at regional, national, and global levels. Farming is but one aspect of a larger, transnational agrifood system that is controlled increasingly by agroindustrial firms. These businesses buy the vast majority of farm produce; they often specify inputs such as feeds, pesticides, and fertilizers; and they also set quality standards for produce and livestock that often demand heavy chemical applications. This is becoming increasingly the case as more and more commodities are produced under contract farming arrangements. Even the environmental aspect of sustainability cannot be understood outside the larger economic context, since the social and economic structure of agriculture (e.g., land tenure, resource allocation and regulation, and the terms of trade) affects environmental quality.

Interactions among the larger environmental, social, and economic systems in which agriculture is situated influence agricultural production and distribution. For example, the increase in the availability of chemical fertilizers and pesticides after World War II was not led by demand by agricultural interests but was a consequence of wartime research and manufacturing. Similarly, the development of pesticides was an unintended byproduct of research conducted on military chemicals. The fact that they were lethal to insects was incidental; they were tested on insects because insects reproduce rapidly, which enabled the acceleration of the research (Duncan 1996). After the war, pesticides were commercialized, and the government encouraged farmers and municipalities to use these compounds. Thus, a key issue that concerns sustainable agriculture activists has little to do with the logic of agriculture itself but with the military-industrial complex that spawned a particular mindset for attacking problems. "Better living through chemistry" was more than just an advertising slogan; by the 1950s, it had become a virtual ideology.

Today, distinctions between agriculture and industry have dissolved even further. Developments in molecular biology directed toward redesigning nature have produced genetically modified organisms that would never have evolved "naturally" in food and agricultural systems. Agricultural production is subordinated to industry through its dependence on industrial inputs and credit. Crops are often no longer produced as final consumer goods but as raw materials for the food industry. Categories within the food and agriculture sector have become increasingly blurred and arbitrary. The potato, for example, is produced by "agriculture," processed into French fries by "industry," and prepared for sale in restaurants through food "service" (Friedmann 1994). Under these circumstances, it makes no sense to cling to the farm as the primary locus of change.

While research into methods of sustainable agriculture has expanded significantly in terms of addressing environmental problems, much more remains to be done to address complex social questions.[3] How does private ownership compare to community ownership in promoting environmentally sound production practices and equitable access to food? Should public funds be used to subsidize the transition to sustainable agriculture? Are objectives of maximizing profit and resource conservation compatible? Questions about land tenure, poverty, and gender equity are either absent, nominally considered, or pushed into the background. Where social issues are acknowledged, such acknowledgments are minimal and often *pro forma*.

From the problems associated with the scientifically based Green Revolution, it is clear that neither science nor new technologies can by themselves solve larger food and agriculture problems. The overarching causes of nonsustainability are not primarily the absence of proper technology or sufficient information about technology. Problems such as inequitable access to resources are plainly outside the scope of the natural-science approaches currently favored by sustainable agriculture. "Neither a lack of technology nor a lack of understanding of ecological processes are standing in the way of sustainable agricultural systems today"; the problem is that most farmers cannot use this knowledge and still survive in the

3. A summary of questions raised through a multidisciplinary conference, "Sustainable Agriculture: Balancing Social, Environmental, and Economic Concerns," is reported in Allen and Van Dusen 1990.

current political and economic structure (Foster and Magdoff 1998: 45). Answering basic questions about agricultural sustainability "requires going beyond narrow—and, frankly sterile—debates on the technologies that might make for a more 'sustainable agriculture' and confronting instead the political and economic forces that have driven farmers into agriculture's present disastrous cul-de-sac" (Clunies-Ross and Hildyard 1992: 7). Attention to *how* agricultural products are produced should be supplemented, as Altieri (1988) proposes, by attention to *what* agricultural products are produced and *for whom*.

Related to the issue of whole systems is the one of causation. Often focused primarily at the level of symptoms rather than causes, little effort has gone into answering basic *why* questions about problems cited by alternative agrifood movements. For example, the advocates of sustainable agriculture promote reduction of pesticide use because pesticides cause groundwater contamination and remain as residues on food. What is less often considered is how and why pesticide use has become so prevalent in agriculture and what the forces are behind this. Alternative agrifood efforts address the more immediate causes of problems. They rarely address their more opaque, structural causes, that is, causes that reflect deeper, systemic operating principles of the food and agriculture system. This limitation tends to naturalize what is social and imposes restrictions on what remedies are possible.

Despite a long history of agrarian struggles in the United States, these struggles have never endured, in part because there were no theoretical explanations for the problems they faced (Friedland et al. 1991). Alternative agrifood movements and programs tend to progress directly from problem description to prescription without the intermediate step of problem explanation. They share with other environment-oriented movements a tendency not to examine the fundamental causes of the problems they confront. For example, Boggs (1986: 220) finds that the German Greens tend to ignore "the structural and material sources of domination that lie at the heart" of environmental crises. This sustains unequal relations of power and refuses to look at ideas that might question existing relations of power and the possibility that the problems are not technological or biological but social. Social problems require social analyses and social solutions.

Framing sustainable agriculture in a natural-science discourse, which excludes social relations, not only ignores social consequences of environmental problems, but leaves unexamined the degree to which environmental problems have social causes. Achieving the goal of environmental preservation is not possible without transforming social institutions and policies. Soil erosion, for example, is a "natural" process but is greatly accelerated by continuous, intensive cultivation practices encouraged by certain agricultural policies. Similarly, declining water tables, common in many agricultural regions, are caused by extensive irrigation, also encouraged by agricultural investment and tax policies. The USDA recognizes that the problems facing rural America are largely due to social, economic, and cultural conditions and, as such, "cannot successfully be addressed solely with the knowledge generated by the biological or agricultural sciences" (USDA 1993a: 48). Still, funding for social issues research in the agrifood system is scarce. Support for social science research in agriculture, which has always been very limited, has actually declined over the past decade (USDA 1993a). In 1987 only one percent of USDA research funds was spent on projects in sociology or anthropology (National Science Foundation 1989).

Overemphasis on natural sciences and farm-level production in alternative agrifood efforts embraces an epistemology that is at once too abstract and overly positivistic. It is overly abstract in that it does not place sufficient emphasis on the social relations through which people produce food and shape the agricultural system. It is too positivistic in that it focuses on what is empirically verifiable and quantifiable without considering the ambiguities inherent in the social relations of production or the quality of life issues important to both producers and consumers. While agriculture's predominant natural-science-based epistemology is essential in preserving or reconstructing the environmental conditions of production in agriculture, by itself it is insufficient to articulate the social relations of production. This focus on the descriptive (rather than the prescriptive) and on observable "things" leads to a reification of social relations in food and agriculture. That is, agricultural research tends to treat social relations as fixed qualities and features of an external, natural world rather than as part of a social world.

Practices of Alternative Agrifood Movements

Practices are the ways in which people organize their labor and apply their ideas to create their world. They constitute a point of connection between abstract structures and processes and concrete events (Chouliaraki and Fairclough 1999). As discussed in Chapter 3, developing alternative institutions and practices has been a central strategy of alternative agrifood movements.

Emergent Practices

Those in the alternative agrifood movement believe that comprehensive changes are needed to develop sustainable and equitable agrifood systems. For example, UC SAREP's strategy for change is broad-based: "In addition to strategies for preserving natural resources and changing production practices, sustainable agriculture requires a commitment to changing public policies, economic institutions, and social values" (UC SAREP 1991: 4). Community food planning and community development are integral to community food security projects. A key step in a community food security project is assessing the food system. The framework that community food security provides may substantially alter urban planning efforts, which have never focused on food provision as such. While cities and counties have departments that address basic needs such as water, housing, health, and transportation, no municipality in the United States has a department of food (Fisher 1997).

While traditional approaches to food and agriculture have tended to separate production and consumption, alternative agrifood approaches often unite moments and agents of production and consumption. There is a clear emphasis on improving understandings and relationships between producers and consumers as seen in alternative agrifood practices. Key features of community food systems include farmers' markets, community-supported agriculture, organic produce, and food-based microenterprises (as described in Feenstra 1997). In addition to bringing farmers and consumers closer together, these approaches offer many other advantages, such as access to fresher, safer food. For example, there is evidence that CSA members eat more fruits and vegetables than do nonmembers. Other benefits of alternative agrifood practices were discussed in Chapter 2.

A primary goal of alternative agrifood practices is the creation of a more equitable agrifood system. For Henderson (1998), for example, a local or regional food system will be environmentally sound and produce healthy food and distribute benefits fairly so that food access will not be based on income: "Every direct purchase from a local farmer becomes an act of fair trade, and every square foot of home garden, every family-owned farm, and every value-adding cooperative becomes a small piece of liberated territory in the struggle for a just and sustainable society." Indeed, some alternative agrifood institutions function as a kind of privatized income-redistribution system. For example, CSAs and farmers' markets can improve the incomes of small farmers because they retain a greater share of the food dollar. Some efforts also transfer income to low-income consumers. For example, a California organization sells expensive, natural foods in health food stores to predominantly affluent consumers. While little of the food is consumed in the low-income community that produces it, the money from its sale is used to support scholarships for people who would otherwise have little chance of going to college. Similarly, another organization produces and sells high-priced organic foods to an elite market, but the purpose of these sales is to generate revenue to support programs for homeless people. Thus, even if they are not actively working toward basic structural changes that would eliminate poverty and homelessness, alternative agrifood institutions are making a difference, day in and day out, to many people who have been marginalized or discarded by the current agrifood system. For example, community supported agriculture can help bring fresh fruits and vegetables to places where they are not usually available, and there are an increasing number of efforts by csas to provide shares or deliver produce to inner-city residents (e.g., the Hartford Food System in Connecticut and Just Foods in New York).

One of the most powerful aspects of these kinds of strategies, however, is less "practical" than conceptual. That is, engaging in alternative agrifood practices can also change participants' consciousness about the agrifood system. "Social practice does not merely 'reflect' a reality which is independent of it; social practice is in an active relationship to reality, and it changes reality" (Fairclough 2001). As awareness is built of the need for change, it simultaneously inspires people to work toward change. For example, the importance of the growth in the organics market lies primarily in the opening it provides for the conscious "defetishization" of food, and for enjoining people to think critically about the food system (Allen 1999). This is also a key value in the fair trade movement. In Raynolds's (2000) analysis of the international "fair trade" approach, she points out that with fair trade, the importance is not the volume of trade—which is extremely small—but in the challenge it presents to exploitative relations in the agrifood system.

Alternative agrifood practices may have effects in ways that are unexpected or out of proportion to what it seems they can actually accomplish given their limited size and distribution. For example, it is possible that alternatives like community-supported agriculture may indeed begin to increase members' interest and engagement in food-system problems and solutions. In our study of csas in California, we found that not only did people improve their eating habits, but that for a few members participation in community-supported agriculture deepened their understanding and activism in the agrifood system. We found evidence that this kind of critical consciousness is present and growing within California agrifood institutions. For example, in one interview, a young, "typical" environmentalist who said he had no position on social justice spoke later in the interview about how treating food as a commodity resulted in putting profits before feeding hungry people. Indeed, a common perspective among those working in California alternative agrifood institutions was that if people began to view food as more than just a commodity, it could lead to political, economic, and social changes in both the food system and in the overall society. According to the president of the board of directors of the CFSC, one of the major strengths of the concepts and activities of community food security is that it gives people a new way to think about food that shows linkages and possibilities for change (Hendrickson 2002). Participation in alternative agrifood practices may get people and communities to think about issues they may never have confronted or considered before, and become effective agents of agrifood system change.

Retention of Dominant Practices

It is odd, then, to observe the extent to which alternative agrifood goals have been operationalized in ways that attempt to remedy agrifood-system problems within existing economic and social arrangements, either through self-provisioning or entrepreneurial development. Alternative agrifood organizations seem to accept many of the structures and parameters of the current agrifood system, and are working more to develop alternatives within that system than to reconfigure the system itself. Perhaps of necessity, these groups are focused more on the day-to-day operations of the business or technical aspects of their work than on formal or informal political activities or broader food-system change. Few interviewees in our study of California alternative agrifood organizations cited strategies that focused on political work or dealing with fundamental issues such as ownership and compensation as important for changing the agrifood system. In their programs and projects, market-based and entrepreneurial activities were predominant. Three-quarters of the organizations engaged exclusively in entrepreneurial activities such as creating niche products or expanding markets, and nearly all organizations engaged in entrepreneurial activities as part of what they did. Only one organization's activities focused primarily on changing the food system through public policy and reforming food-assistance entitlement programs. While antihunger programs have been necessary precisely because the market has failed to ensure food security, many alternative agrifood activists see the market as essential for achieving food security.

Alternative agrifood efforts often place greater emphasis on entrepreneurial than on entitlement approaches (see Allen 1999). For example, the projects funded in the first two years of the USDA Community Food Projects Grants emphasized entrepreneurial-type projects, while working with existing federal food programs was on the agenda of only one of the projects funded. Only in a few instances were policy efforts undertaken or attempts made to connect community food security with traditional food programs for the poor. Of the thirty-one projects funded, only four included a food-policy or food-system planning effort focused on food security. Of the four projects that focused on food-policy or food-system planning, all were concentrated on local actors; none addressed the federal-level policies that significantly shape food security in local communities. Perhaps to address this situation, the 2002 USDA Community Food Projects program request for proposals points out that grant applicants should also recognize the role played by food and nutrition-assistance programs administered by the USDA. While the projects are directed toward meeting the food needs of low-income people, they do so with the assumption that this can be achieved through local, market-based initiatives. Providing food for people outside the market system was featured in very few of the projects. Indeed, one of the strengths of the community food security movement in getting its legislation passed through Congress was that it shared the discursive basis of self-reliance and entrepreneurship with the Republican majority.

As mentioned, a key hope and claim for alternative agrifood practices is that they will be more equitable than those of the globalized, industrialized agrifood system. For example, Friedmann (1994: 272) posits that the continued industrialization of the agrifood system will lead to a situation in which food consumers are "differentiated by class, rather than nation or cultural region." Yet food consumption has always been differentiated by class. Historically, the food system was two-tiered, with cheap, mass-produced foods to meet basic needs of the masses on the one hand, and highly elaborated, individually tailored goods produced for a powerful fraction of the population on the other (Fine and Leopold 1993). Distinctions between luxury and basic foods began to disappear to some degree as the industrialization of the food system expanded the range of products available,

provided safe and high-volume food preservation, packaged food in con-
venient amounts, and allowed for the sourcing and transportation of food
over long distances. Prior to the development of the cash economy, the
consumption of certain foods was determined by traditional categories of
status and rank; as consumption has been determined by ability to pay, these
barriers have broken down (Winson 1993). If anything, the industrialized
food system has reduced class differences in food consumption, a leveling
that some alternative agrifood practices may unbalance.

The promotion of social justice through these practices has not been
straightforward. Community food security focuses on the food needs of low-
income people on the one hand and the need for local food production, sustainable agriculture, and local/regional food systems on the other (Gottlieb
and Joseph 1997: 10). Production and consumption goals may not always
be compatible and may even contradict one another. Marketing strategies
in community food security are often directed toward increasing demand
and outlets for local produce, providing security of markets for local agricultural producers, and creating product differentiation based on region of
production. Local food-systems projects based on provincialism and local
brand-name recognition may actually serve the status needs of the privileged
more than the material needs of the poor. This is an issue with which the
CFSC is actively struggling. Should community food security be defined
mainly as a food security strategy for low-income people? Or should it be
primarily concerned with developing sustainable approaches to local food
production and distribution with less regard for the needs of low-income
populations? There may be contradictions in trying to do both of these.

Alternative agrifood practices may actually contribute to the class differentiation of the agrifood system. As fresh fruits and vegetables become
"branded" by place or differentiated by method of production, prices
increase. For example, organic foods carry a price premium of as much as
300 percent. One study found that price premiums for organic produce
ranged from 40 percent to 175 percent (Thompson and Kidwell 1998). The
question of whether price premiums reflect higher production costs or a
type of market "rent" remains unresolved. The Organic Farming Research
Foundation explains that organic food costs more because, in order to meet
stricter regulations, producing, harvesting, transporting, and storing it is
more labor intensive and farming is on a smaller scale (Organic Farming
Research Foundation n.d.). They further point out that if the indirect costs
of conventionally produced food (e.g., water pollution, soil erosion, and
health care for those who work on farms) were included in the price of such

food, organic foods would cost the same or possibly be cheaper.

Nonetheless, for now organic food usually does cost more than "conventional" food. This can be an enormous obstacle for low-income people, who may already be paying a price premium for their food, since food prices tend to be higher in stores in low-income neighborhoods. In addition, since nutritional needs are the same for people regardless of income (i.e., they do not increase as incomes increase), people with low incomes must spend a larger percentage of their incomes on food than do those with middle or high incomes. A single parent earning minimum wage will need to spend much more of his or her earnings on food than will a single parent earning an executive-level salary. A study conducted in Los Angeles found that a low-income family spends 36 percent of its annual income on food as compared to 12 percent for a middle-income family (Ashman et al. 1993). Organic food may be beyond the reach of the working poor. This is an issue that some in the organic industry tend to ignore. For example, one organics industry executive said, "The mainstream consumer is voting with her pocketbook for a better world for her and her children. That consumer doesn't give a darn about money when it comes to what she's feeding her children" (Nachman-Hunt 2002: 40). This somewhat elitist approach does not consider that "she" might not have that kind of discretionary income.

In the California food-systems projects of the UC SAREP program, the majority of the projects profiled in the report *Community Food Systems in California* (Feenstra and Campbell 1998) were focused on developing local or regional food systems by increasing demand and outlets for local produce. In these thirteen projects, the central focus is less on the food needs of low-income people than on the "well-being" of a region. Food distribution in this context means "marketing" rather than "access." While a few are oriented toward food access for low-income people and the development of food policy councils, most are directed toward providing security of markets for local agricultural producers and creating product differentiation based on region of production. This approach resembles the "appellation" marketing approach of the wine industry.[4]

Sustainable agriculture and community food security efforts such as local production and marketing need to be evaluated for their ability to increase

---

4. In the wine industry, *appellations d'origine* are used as geographical indications of provenance that combine the natural environment of a region and certain production standards (Bell and Valentine 1997). They function as a sort of a status brand or certification of quality and have begun to be applied in other food industries as well.

social equity. Entrepreneurial food security projects may not be able to address structural inequalities that produce hunger in impoverished communities. Liz Tuckermanty, director of the USDA Community Food Projects Program, points out that the economic development aspect of the projects is a weak link of the community food security approach (Tuckermanty 2002). For example, food-oriented jobs are traditionally low-paying, so it is difficult to see how food-based microenterprises can be successful at lifting people out of poverty.

Alternative agrifood practices may also inadvertently reinscribe class privilege. In Cuba's urban agriculture system, for example, local production and marketing has increased the aggregate consumption of fruits and vegetables in Havana, but this improvement is not equally accessible to those at all income levels. Prices are set by supply and demand, and access is based on ability to pay (Fuster 1998), even in this putatively nonmarket economy. Additionally, in the United States, farmers' markets are not frequented by the lowest-income consumers. Less than 25 percent of food-stamp recipients reported shopping at a farmers' market, and food-stamp redemptions at farmers' markets in 1988 accounted for only 0.02 percent of overall food-stamp redemptions (Kantor 2001). This participation dropped even lower when many states adopted electronic Benefits Transfer (EBT) systems for food stamps because the electric power and land-line telephone access required for EBT redemptions are often not available at farmers' market locations.

The same kinds of class-based restrictions may apply to CSAS. For example, a basic tenet of community-supported agriculture is that shareholders share risks with farmers. A key advantage for farmers is that they receive income at the beginning of the planting season rather than after crops are harvested. But this approach constrains the participation of the poor, who do not have the cash reserves that would enable them to make large up-front, or even monthly, payments. CSAS offer a type of futures market in which people invest based on the presumption that food will be produced and delivered, but this is not assured. In fact, CSAS have been prohibited from accepting food stamps because by providing before-harvest payments to farmers, members are actually speculating on the crop rather than purchasing food. While there are examples of CSA efforts to provide shares or deliver produce to inner-city residents, this is made possible only because of external funding sources, not market forces.[5]

---

5. The fact remains that there is a contradiction between making food affordable and providing a decent return for the farm unit in the absence of public subsidy. Charity efforts cannot overcome this contradiction in the long run.

Furthermore, since consumers cannot get all of their food from a CSA, it means they have additional shopping tasks. The idea of getting everything in the same place is what gave rise to the supermarket in the first place, and for most working people time has become even scarcer. Additional shopping trips can place a burden on working people, many of whom are working longer hours and spending more time commuting. There is also the practical problem of the rapid deterioration of quality and nutritional value of fresh produce that is not refrigerated, and the issues of storing and preparing the food. Processing, preserving, and packaging activities do not disappear, they are simply displaced to a different time, space, and labor relation. Somebody still has to do the work—it is simply reembedded in the home, the domain of women's labor, particularly in regard to food.

In this context, the idea of community-supported agriculture seems anachronistic for all but the most privileged. While CSAs are an excellent alternative for those who can afford them, there is evidence that CSA shareholders tend to be a rather select crowd. Studies have found that CSA members tend to be predominantly upper-income, highly educated European Americans (Cohen et al. 1997; Festing 1997). In our study of CSAs on the Central Coast of California (see Perez et al. 2003), we found that the vast majority (90 percent) were European-American, while European Americans made up only 51 percent of the people in the study area. CSA shareholders had higher incomes than average for the counties, and they were more highly educated. Eighty-one percent were college educated. In a 1992 survey of California CSA members, only 10 percent of the farms had members who were people of color (constituting only 5 percent of the overall membership) (Lawson 1997). A study to explore the significance of class in CSAs examined occupation, education, and income of members and found that that 56 percent of CSA members held postgraduate degrees (Hinrichs and Kremer 1998). Interestingly, even CSA projects that actively sought to increase low-income participation through financial subsidies ended up attracting low-income educated professionals rather than the working class or the traditionally poor (Hinrichs and Kremer 2002). Further, while evidence exists that CSA members eat more dark-green and yellow fruits and vegetables, fiber, and vitamin A than nonmembers (Cohen et al. 1997), they do not come from populations with a priori low intakes of these nutrients.

How and to what extent are alternative agrifood practices beginning to reshape practices of the agrifood system? Where they present an alternative to the dominant agricultural paradigm is not in the idea of a market economy but in the geographical and class boundaries of the "consumers" they seek. In many alternative agrifood practices it appears that what is to be secured are markets for producers rather than food access, *per se*. While the dominant paradigm pushes overseas markets, community food security and sustainable agriculture are focused on local production for local markets. While this may ensure that people can have more control over the source and quality of their food, it does not address basic questions of equity and access. To a large extent, there is a continuation of the neoliberal orientation of the conventional agrifood system embedded in alternative agrifood efforts, particularly in their emphasis on creating opportunities for producers and consumers to have different options and make different choices. A dominant orientation of the alternative agrifood movement is toward developing alternative businesses that would allow people to acquire fresher, healthier food or help small farmers become or remain viable. For many alternative agrifood efforts, changing the food system means increasing the diversity of sustainable markets and ensuring that consumers have more choice, rather than making deep structural changes in the food system that could reconfigure who it is who gets to make which kinds of food choices. Yet, as Hinrichs (2000) cautions, we need to closely examine the new economic systems arising in alternative agriculture movements in order to understand what they mean for the movement and for those involved in their operations, rather than assuming that they are meeting goals for change in the food system. In her analysis of direct markets, she observes that direct marketing from farmer to consumer does not necessarily or fundamentally challenge the commodification of food. Thus, market-based alternative agrifood practices may be limited in their potential to transform the agrifood system in the direction of social justice.

Through the alternative agrifood movement progress has been made in developing programs, legitimizing the concepts of sustainability and community food security, expanding research agendas and approaches, and developing alternative food production and distribution practices. Yet these emergent forms do not yet deeply interrogate the conditions that retard or enable the achievement of agricultural sustainability and food security.

5	reflections on ideologies embedded in
	alternative agrifood movements

The combination of the development of alternative agrifood institutions and the integration of sustainable agriculture and community food security into dominant agrifood institutions has begun to make significant changes in agrifood discourses and practices. Concepts of agricultural sustainability and community food security are becoming increasingly familiar and better developed, programs in these areas have been established in traditional agrifood institutions such as the USDA and land-grant universities, and alternative agrifood institutions such as community-supported agriculture are proliferating throughout the country. Furthermore, most of this change has happened in the past decade. This development and integration is the result of many forces, including the crucial nature of food and its omnipresence in our daily lives. Another reason for this rapid growth is that the alternative agrifood movement challenges the conventional system, but not to such a degree that it poses a serious threat. This allows the movements entry into institutions from which they would otherwise be excluded and allows for the further growth of alternative agrifood institutions. This circumscription of challenge can be seen as smart, practical politics.

While this approach has contributed to the acceptance and progress of alternative agrifood movements, it can also have neutralizing or contradictory effects. That is, the problems and solutions that tend to be articulated are those which can be addressed within the framework of traditional epistemologies and practices—including those which have contributed to agrifood problems in the first place. Often the problems amenable to resolution by traditional approaches and institutions predominate, such as research in new pest management techniques, development of community gardens, or improvement of marketing systems. In this way, despite conceptual and methodological departures from the path of traditional agrifood research, alternative agrifood programs have tended to reproduce key frameworks and foci of the conventional system. This accommodationist approach can also take conflicting issues such as social and environmental justice out of the framework. For example, while alternative agrifood movements deal directly with the most glaring manifestation of social injustice—lack of access to food—this is generally approached within a neoliberal framework that precludes an in-depth examination of the fundamental causes of food insecurity. This is an illustration of Hajer's (1995) claim that political conflict is hidden in how problems are defined, which aspects of social reality are included, and which are left out.

This is not necessarily intentional, but instead results from uncontested, imbricated—"normal"—ways of thinking, speaking, and acting in the world. These unexamined ideas and practices are rooted in assumptions and ideologies that shape the discourses and practices of alternative agrifood movements. It is worth examining more closely the ideological orientations embedded in alternative agrifood movements to see where they may facilitate or obstruct the realization of their articulated goals. Some of these ideologies may be inadvertently reproducing aspects of the dominant agrifood system, limiting progress toward agricultural sustainability and food security. This is because the institutional practices people draw upon without thinking can embody assumptions that directly or indirectly legitimize existing power relations (Fairclough 2001).

Moving the agrifood system toward greater social and environmental justice requires reflecting upon the assumptions and ideological formations embedded in alternative agrifood movements that may limit their advancement, particularly toward their goal of social equity. This reflection is the purpose of this chapter. In this I take a critical approach in the sense used by Fairclough (1994), that is, to illustrate how discourse is shaped by the ideologies and relations of power. As context, I first discuss the role of ideology and hegemony in shaping social perspectives and practices. I then review five ideological orientations that I see as constitutive of the discourses and practices of alternative agrifood movements, but which may also constrain progress toward social justice. These ideologies are agrarianism and farm centrism, individualism and self-reliance, economic liberalism, ideologies of class and merit, and environmental fetishism. At first glance, these may seem like progressive approaches, and they do contain liberatory aspects. At the same time, they can also function to legitimate current social structures, reinscribing asymmetric relations of power and privilege in the agrifood system.

Ideology and Hegemony

I have posited that social movements are constituted in large part by their discourses. Ideologies are embedded in these discourses and shape the range of alternatives imagined and developed by the social movements. In his book, *The Sociology of Culture*, Raymond Williams (1981) explains that ideology is generally used in two ways. One meaning of ideology is the formal, conscious beliefs or general principles of a particular social group, such as a religious organization. The second meaning, and the one I explore in this chapter, is the worldview or general perspective of a social group. Ideologies are significations and constructions of reality (Fairclough 1994). They form the unquestioned assumptions that organize how we think and live our lives.

The ideologies embedded in discursive practices are most powerful when they become naturalized and achieve the status of common sense. Such ideologies become hegemonic—a concept developed by Antonio Gramsci— in that they are totalizing and omnipresent. Williams (1980) explains the character of hegemony, saying that it "saturates the society to such an extent" that it "constitutes the substance and limit of common sense" for most people in a society. Hegemonic ideologies determine our assignments of energy and shape our ordinary understanding of the world. As such, they constitute a sense of reality for most people in society, a reality beyond which it is very difficult for individuals or organizations to move. In the United States, pervasive and long-standing ideological formulations include the economic, such as the notions of the free market, individualism, and equal opportunity; the political, such as that of representative democracy; the epistemological, such as science; and the cultural, such as agrarianism, patriarchy, white superiority, and the formulation of the "undeserving poor." Some of these ideologies reinforce class or group cohesion and legitimate domination by powerful groups by fostering the acquiescence of subordinate groups or classes.

Ideologies do not require manipulation by powerful agents to become hegemonic. Rather, ideologies seem to be confirmed over and over again through experiences of daily life and therefore become "naturalized." The way the world "is" insidiously translates and mutates into how the world "can only be" or "should be." As a result, many values, cultural traditions, and political economic structures remain largely unquestioned. This process is recursive in that it begins a new iteration of assumptions that shape the range of future possibilities for solutions to social problems.

Furthermore, the relationship between the material and the ideational is dialectical. Historical patterns of access to and exclusion from resources shape cultural understandings of rights, property relations, and entitlements (Moore 1996). These cultural understandings in turn affect the social and material structure of society. Culture shapes power, and power shapes culture. Williams (1973: 9) calls this operating process the "selective tradition," namely, that which is always passed off as *the* tradition of the dominant culture. "But always selectivity is the point; the way in which from a whole possible area of past and present, certain meanings and practices are chosen for emphasis, certain other meanings and practices are neglected and excluded." Dominant cultures are constituted by the central system of practices, meanings, and values that are not abstract but are embodied in our daily lives. These practices, meanings, and values are hegemonic in the sense that they frame our expectations, direct our energies and efforts, and constitute our ordinary understanding of the world. Ideologies are crucial to understand because they are built into discourses and practices and thus contribute to the production, reproduction, or transformation of social relations. Ideologies may either promote or obstruct progress toward social justice and environmental sustainability. As a framework for discussing ideologies in alternative agrifood movements, I use Williams's categories of residual, dominant, and emergent.

Residual beliefs and practices are derived from an earlier stage of society and may reflect a social formation different from the present's. They may be appropriated by the dominant ideology in order to give itself authority through tradition. The residual ideology I discuss here is that of agrarianism and farm centrism. Dominant ideologies are those which prevail in and organize a social formation. These ideologies are produced and reproduced as material practices as they become absorbed into the everyday consciousness of both dominant and subordinate groups in the way people make sense of the world. They become incorporated into our unconscious assumptions, habits, and commitments. I discuss three formulations inherited from the dominant social system: individualism, economic liberalism, and ideologies of entitlement and merit. Emergent elements are those which are substantially alternative to dominant ones, not merely new elements of the dominant culture. Into this category I put the emphasis on the environment in alternative agrifood movements.

Agrarianism and Farm Centrism

A prevalent viewpoint within alternative agrifood movements is that a sustainable and equitable agrifood economy can and should be based upon a family-farm agrarian structure. The focus on the family farm in alternative agrifood movements is grounded in a long-standing tradition of agrarian ideology. Some, such as Buttel (1992), consider the sustainable agriculture movement to be the successor to the 1970s family farm activism. Agitation during this period, in turn, drew upon the agrarian populist movements of the previous century. Like the populists of the 1800s, alternative agrifood advocates oppose the industrialization of agriculture and loss of market control and advocate traditional rural values of independence, hard work, and strong family ties. Agrarian ideology is constructed around the independent land-owning producer, who is considered the foundation and bearer of American democracy. The family farm, based on the notion of the virtuous, hardworking farmer, has been the basis of American farm policy since the 1800s and continues to live in the hearts and minds of both rural and urban Americans.

Alternative agrifood movements have tended to accept traditional agrarian notions of the ideal of the family farm and traditional rural values. Agrarianism is part of the American national identity and shapes people's images of farmers as well as farmers' images of themselves (Danbom 1997). Family farms are viewed both as essential to viable rural communities and as the primary agents for developing an improved agrifood system. An illustration of the hegemony of agrarian ideology is the frequent reduction of agricultural sustainability to farming. Even in an article calling for more attention to policy issues in sustainable agriculture—and in a section on demography and sociology—the focus remains exclusively on farmers and farming: "How many sustainable farmers are there in the United States? How are such farms distributed geographically? What methods do they use? How do these methods vary from one region to another? Why? Are there sustainable farms in all regions? If not, why not? What role does the farm family play in the success of sustainable farms?" (Youngberg et al. 1993: 314). This article does not discuss the other dimensions of the food system or its host of other actors, such as consumers, farmworkers, retailers, and researchers. This kind of farm centrism is widespread in alternative agrifood movements.

Agrarianism upholds a belief in the moral and economic primacy of farming over other occupations and ways of producing (Fink 1992). For example, alternative agrifood advocates consistently point to the reduced share of the food dollar that accrues to farmers, overlooking the fact that farmers are generally better off economically than farmworkers or food-processing workers. Even in the new economic forms designed to forge a closer relationship between producers and consumers, farming tends to be both romanticized and privileged over other forms of labor. For example, in some visions of the ideal model of community-supported agriculture, members share in the labor of farming, yet farmers are not expected to reciprocate by helping the members perform their regular jobs. Most middle- and low-income people, already overwhelmed with the demands of productive and reproductive labor in their own jobs and households, have precious little time for extra activities. This orientation is an example of a rarely acknowledged farm centrism in the alternative agrifood movement.

Farmers receive salutary treatment for their economic troubles that other sectors of society do not. For example, during the same period in which the country was seized by a fervor directed at moving poor women off welfare and into the workforce, the government simultaneously increased subsidies to farmers. Bourdieu and Wacquant (1992) point out that some social practices take on more value and status than others, and that knowledge of and access to those practices puts some people in more powerful positions. For farmers, when times get tough, subsidies are simply reinstated. For example, when crop prices began to drop because of the 1997 collapse of the Asian market, policymakers set up a $7.1 billion bailout package for producers (Belsie 1998). The power of agrarian ideology is such that few politicians would have dared to suggest that farmers likewise go from "welfare to work." Farmers are discursively juxtaposed to the poor at the top administrative levels of the USDA. In congressional testimony on the USDA's work in sustainable agriculture, the assistant secretary for science and education, referring to USAID's mission to "increase the income of the poor majority and expand the availability and consumption of food while maintaining and enhancing the natural resource base," said, "If you were to substitute the words 'American farmer' for 'poor majority' you would have a valid mission statement for the USDA's sustainable agriculture program" (Senate 1992). In this formulation, the poor in the American agrifood system would be written out of the equation.

Farming has always been a privileged category in U.S. policy, and agrarian ideology has enabled a certain naturalization of the need for existing agricultural policy. For example, labor law has always provided fewer protections and recourse for agricultural workers than for workers in any other industry. Acknowledging that it is unsurprising that the USDA does not regulate farmers, since they are its primary constituency, Merrigan (1997) points out that this treatment extends into other federal agencies such as the Environmental Protection Agency. She writes that in the twenty-three years since the passage of the Endangered Species Act, only one landowner has ever been successfully prosecuted under this law. Despite the special ways in which farmers have been treated in federal policy, agrarian ideology nonetheless is generally politically conservative. For example, agrarian populism promotes a disdain for "outside" forms of power and authority such as governments and experts (Stock 1996). The particular experiences of rural life in America, such as isolation and being exploited by industrialization, along with a gun culture and militantly self-defensive communities, has produced collective acts of violence and movements based on hatred and white male supremacy (ibid.). Stock points out that the Ku Klux Klan and the John Birch society were products of rural America.

Agrarian ideology contains elements such as community and cooperation that resonate with the progressive character of alternative agrifood movements. At the same time, agrarianism also embodies some regressive values such as racism and sexism. Although agrarian populism has in the past emphasized grassroots democracy, organizing and solidarity were largely based upon ethnic homogeneity and a traditional gender division of labor (Naples 1994). Fink (1992: 196) characterizes the exclusionary nature of agrarian ideology, stating that it has been "a white male vision that has failed to consider the full human integrity of other persons." She points out that agrarianism is a gendered ideology that projects different ideals for men and women. Women have been expected to support the farm, men, and children ahead of their own needs or aspirations (which were not even acknowledged categories). Focused on the nuclear family and the male farmer, agrarian ideology embodies traditional gendered roles and can pose a roadblock to raising issues of gender equality in alternative agrifood movements.

The neopopulist agrarian call for a return to traditional rural values also elides the thorny issue of the historical congruence between racism and agrarian populism in the rural United States. Stock (1996) suggests that since the development of agrarian America required the dispossession of people of color from their land and in the South was based on the enslavement of people of color, rural Americans feared retribution from them. This, combined with an absence of people of color in many rural communities, enabled an intolerance and sometimes scapegoating of nonwhites. Rural America has been a major arena for the subjugation of people of color in more recent times as well. For example, in California after World War I, advocates of a family farm agrarian structure lobbied to exclude ethnic minorities from the fields in order to provide opportunities for soldiers to establish farms (Pisani 1984). A parallel can be drawn here with the establishment of the American slave economy. In the first years of colonial settlement in America, blacks could own land. Not until white indentured servants became free and sought their own land was the category of race constructed as a barrier to land ownership. Agrarian populists were little concerned about the subsequent process of the separation of African-American farmers from their land, despite its extent. In 1910, black farmers owned over 15 million acres in the South; by 1970, they owned 6 million (Healy and Short 1981). In contrast, there was a strong resurgence of agrarian populism during the farm crisis of the 1980s when those affected were primarily European Americans. Although the farm crisis of the 1980s affected mostly large, highly capitalized farms,[1] publicity about and responses to the situation were cloaked in a yeoman farmer agrarian rhetoric. Landholding and race have always been connected in American agriculture. From the beginning, Thomas Jefferson's ideal of civic virtue depended upon agriculture being the primary occupation of Americans; this in turn for him required that America needed to continue to expand into "vacant" lands (Kemmis 1990). Obviously, this expansion depended upon the subjugation and extermination of the Native Americans who hap-

---

1. In the 1970s there was a surge in export demand caused by increasing income in the oil-producing countries, Soviet grain shortfalls, and poor world grain harvests in the early part of the decade. By 1973 commodity prices were increasing dramatically and stocks were disappearing (McCalla and Learn 1985), and farmers were encouraged to expand production. Then, between 1974 and 1977 farm product prices stabilized or declined and input prices rose dramatically (Cochrane 1979), creating a cost-price squeeze. In the twenty-year period prior to 1979, fertilizer costs increased 80 percent, gasoline costs increased 300 percent, machinery and equipment costs increased 120 percent, pesticide costs increased 250 percent, and farmers paid millions in excessive rents because input prices were set above value (Havens 1986).

pened to live on these lands. This is not to say that those who endorse agrarian populism are racist. Certainly the Southern Tenant Farmers Union did not embrace racism. My objective here is to call attention to the connections among rural values, property, and race that exist within the agrarian frame of alternative agrifood movements.

Agrarian ideology is not only culturally but also economically conservative. American populism was developed and promoted by those who had dominated economic life until their position was diminished by the growth of industry (Rochester 1943). The agrarian populists of the late 1880s were commercial producers who willingly participated in the world agricultural economy (Danbom 1997). What they rebelled against was the distribution of the benefits of that economy, believing that a greater share of profits should go to farmers, because—in their view—farmers were the primary producers of wealth. The populists were generally uninterested in the problems of sharecroppers and farmworkers and did not develop alliances with these groups. This is because agrarian ideology takes present forms of property ownership and agricultural markets as givens instead of as social constructions. While agrarian ideology is critical of dominant economic interests and agricultural trends, it nevertheless assumes the legitimacy of private property and current land-tenure patterns, seeking to perfect rather than fundamentally transform this system (Buttel 1980).

While these regressive aspects of agrarian ideology have not been embraced by alternative agrifood movements, neither have they been addressed or explicitly rejected. For example, the Asilomar Declaration for Sustainable Agriculture, produced through an intensive meeting of U.S. sustainable agriculture experts, states: "The continuation of traditional values and farming wisdom depends on a stable, multi-generational population" (Committee for Sustainable Agriculture 1990). This implies that current rural values—which have often included the patriarchal family and Christian religious beliefs—are ideals sustainable agriculture advocates should promote and preserve. Many in the movements see the family farm as the ideal organizational structure for sustainable agriculture, but they generally leave unexamined the inequitable social relations that have often been part of the family farm structure. Agrarian ideology sets farmers up on a pedestal and emphasizes their bourgeois allegiances. This makes it difficult for advocates of agrarianism to see any class and power circumstances farmers may share with labor, minorities, or the poor. This is a type of essentialism and individualism that, perhaps inadvertently, seems to be valorized in alternative agrifood movements, to the detriment of the more cooperative values these movements espouse.

Individualism and Self-Reliance

In the United States, the individual pursuit of self-interest is believed to produce the optimal social good. In no other country is the notion that people can succeed if they just work hard enough so deeply inscribed. It is not surprising, then, that in the 1990s discourses of rights and entitlements were replaced by neoliberal arguments about individual responsibility. This theme of self-reliance, closely tied to the value and potential of individual choice, is prevalent in alternative agrifood movements.

For example, the community food security movement developing during the 1990s carried themes of self-reliance and individual responsibility (Allen 1999). According to Andy Fisher (1997), the executive director of the CFSC, "Community food security projects emphasize building individuals' abilities to provide for their own food needs rather than encouraging dependence on outside sources such as food banks or public benefits." Community food security activists want to reduce people's vulnerability to the vagaries of food program changes and create opportunities for low-income people to define and develop food security for themselves.

In so doing, however, community food security advocates have tended to conflate emergency assistance and government food security programs into a negatively charged category of "dependence." This conflation of charity and entitlement programs is not one used by the hungry themselves, however. In a study of women at risk of food insecurity, their perceptions of "normal channels" for acquiring food included cash purchase, food stamps, WIC coupons, and school meals; the socially unacceptable channels were food pantries and the charity of friends and family (Radimer et al. 1992). The image of the undeserving poor is so powerful that a liberal nongovernmental organization felt impelled to include a section on its food security web page explaining that having the right to food would not make people lazy (Food First Information and Action Network 1997).

There are also vulnerabilities in using the language of self-reliance and empowerment. For example, a study of local food projects in the United Kingdom found that this type of rhetoric tended to locate both the problems and solutions as belonging to the poor, obscuring the need for fundamental change (Dowler and Caraher 2003). They write, "The problems of inequality are on such a scale, and their health and food dimensions so structurally based, that one could question the likelihood of food projects achieving positive outcomes, particularly those located in the realm of individual behavior" (p. 60).

In general, alternative agrifood organizations have framed their programs in terms of the rights of consumers to choose alternatives, rather than in terms of their rights as people to have enough to eat (Allen and Kovach 2000). For example, the theme of choice was prominent in what California alternative agrifood institution leaders considered to be the most promising solutions to food-system problems. Respondents were encouraged to let their answers range beyond the strategies in which their organizations were engaged and to think about the broader issues. The majority of strategies suggested were oriented toward "choice," such as providing more marketing outlets like farmers' markets, educating people about nutrition so they could make better food-purchasing decisions, and getting people to pay the true cost of food (thereby helping farmers be profitable). This is in accordance with Clancy's (1997) call for a food system in which people have more choices, a menu of choices that people themselves can help to create by being active in the development of markets rather than passively accepting the present food market.

This emphasis on choice is somewhat limiting, however. Even if an economic arrangement based on choice is formally noncoercive, desperate economic need is in effect a form of coercion. For those who lack economic means in a market system based on choice, the distinction between coercion and choice becomes meaningless. As Lappé and Schurman (1988: 41) ask, "When one is hungry, how many choices are voluntary?" Nonetheless, in the United States the poor are sometimes even blamed and villainized for their poverty. For example, popular ideologies of the "undeserving poor" are based on the premise that poor people have bad values and habits and that if they were to change those values and habits they would not be poor (Gans 1995). Cuts to food stamp programs were justified as necessary to end cycles of dependency and to stop undercutting the work ethic by providing "free food." The problem with institutions such as emergency food programs (e.g., food banks and soup kitchens) is not that they create "dependence," but that they perpetuate the perception that food insecurity is being addressed, deflecting attention away from any government responsibility for the poor (Poppendieck 1997). The neoconservative critique of welfare misses any viewpoint that wages are too low (Mills 1996) or that entitlements, whether private or public, can make self-reliance possible.

One result of the dominance of ideologies of individualism is that the poor are not only deprived materially but also demoralized politically and psychologically. While in the 1960s the hungry marched on Washington and sat in at the USDA, today the hungry dejectedly and patiently wait in lines (Poppendieck 1995). Even the poor blame themselves for their situations, as revealed in *The Hidden Injuries of Class* (Sennett and Cobb 1972). The self-blame and demoralization become even more potent as the poor become poorer while the richest layers of society continue to accumulate wealth. Between 1989 and 1995 income for the poorest fifth of Americans decreased 2 percent but increased 43 percent for the wealthiest 5 percent—even without including capital gains income (Center on Budget and Policy Priorities 1996b).

Farmers are also affected by this ideological straitjacket of self-reliance. Adams (1995) discusses how the ideology of individual entrepreneurship was so powerful that farm men and women blamed their problems during the farm crisis of the mid-1980s on their own individual choices—even when they simultaneously understood their situation as stemming from larger political economic structures and exacerbated by government policies that favored expansion and exports. This type of individualistic orientation has not always been part of the agrarian perspective, however. According to Adams (1995), as a means to discredit socialism, large agribusinesses and the Farm Bureau mounted an intensive ideological campaign during the Cold War to transform farmers' existing agrarian individualism, in which household and community were paramount, to entrepreneurial individualism, in which profits were paramount.

The power of ideologies of individualism and self-reliance is a perfect example of hegemony. The effect of hegemony is that it shapes people's perceptions, cognitions, and preferences in such a way that they "accept their role in the existing order of things, either because they can see or imagine no alternative to it, or because they see it as natural and unchangeable, or because they see it as divinely ordained or beneficial (Lukes 1975: 24, cited in Hall 1982). The perception of the poor as undeserving exists even in the liberal enclaves of alternative agrifood movements. For example, one organic foods retailer in Santa Cruz, California, attempted to mitigate the high cost of organic food by offering a discount to customers paying with government-subsidized food coupons. Some other customers were so angered by what they saw as giving the poor an undeserved break that they threatened to boycott the store. Perhaps one reason that ideologies of individualism are popular is that if social problems are treated as individual rather than social, everyone else can be absolved of complicity in contributing to or not helping to solve social problems. This type of extraction of social relations from the realm of the political is the hallmark of economic liberalism.

Economic Liberalism

According to Barham (1997), social movements around alternative agriculture attempt to balance the "laws" of the market on the one hand with human and environmental needs on the other. In Barham's study, leaders of the sustainable agriculture movement in France consistently criticized the market as incompatible with meeting human and environmental needs. In the United States, however, alternative agrifood movements have for the most part accepted the basic structures and operating principles of the dominant economic ideology of economic liberalism. This includes embracing production for profit, believing in the free market, and championing entrepreneurship. The ideology of economic liberalism is based on a system in which land and labor are commodified and a "free" market functions to facilitate capital accumulation for certain groups in society. This ideology, in which social goals are subordinated to profit maximization, has achieved almost an archetypal form in this country.[2]

The American agrifood system has always emphasized high production and profit, and these emphases are often reproduced in sustainable agriculture programs. For example, the USDA (1988) began its sustainable agriculture program (then called LISA—low-input sustainable agriculture), by stating that "LISA helps keep farmers profitable." In the USDA's first brochure on its sustainable agriculture program, "10 Guiding Principles of Low-Input, or Sustainable, Agriculture," was the statement: "If a method of farming is not profitable, it cannot be sustainable" (USDA 1988). The subtitle to the USDA's 1997 pamphlet on sustainable agriculture was "Ways to Enhance Profits, Protect the Environment, and Improve Quality of Life."

The focus on production and profits has been echoed in the scientific literature on sustainable agriculture (see, e.g., Edwards 1990a, Francis 1988, and NRC 1989). Ruttan (1988) emphasizes that enhanced productivity must remain a key factor in any definition of sustainability. The National Academy of Sciences states, "Successful alternative farmers do what all good managers do—they apply management skills and information to reduce costs, improve efficiency, and maintain production levels" (NRC 1989). The inaugural editorial of the *Journal of Sustainable Agriculture* declares that "sustainable agriculture is a system in which resources are kept in balance with their use. . . . Production, profits, and incentives still retain their impor-

---

2. O'Connor (1998) points out that the wage form of labor and commodity form of need satisfaction are much more developed in the United States than in most other countries.

tance" (Poincelot 1990: 1). For Pfeffer (1992: 10), there is no evidence for or inherent reason to believe that the sustainable agriculture movement will "significantly challenge the economic prerogatives of agribusiness." Rather than oppose the political, macroeconomic, and structural context that defines profit and economic efficiency in agriculture, those in alternative agrifood movements tend to accept and work within this framework.

Compliance with this system is required for agrifood enterprises—both conventional and alternative—in order to flourish in the market economy. The organic foods market, developed by committed food producers, retailers, and consumers, illustrates this process. As the market for organic foods expands, large-scale agribusiness is entering the industry, either through converting small sections of their large farms to organic production (thereby competing directly with small organic producers) or through contracts with or purchase of successful organic farms. Many in the organics industry are predicting that growth in organics will lead to a drop in the premiums for organic foods (Nachman-Hunt 2002). Lower prices could hurt the viability of small-scale organic farms. Through this process of growth, in some areas of California organic agriculture has come to resemble conventional agriculture in its dependence on migrant wage labor, use of monocropping, and production for contract (Buck et al. 1997). Small-scale organic food businesses are also growing rapidly or being bought out by agrifood conglomerates. Organic food producers, processors, retailers, and distributors must seek larger and larger market shares as they are inevitably driven by growth imperatives. Although many of the early participants in the organic foods movement are unhappy with these developments, they do not see how to change this trajectory (Vos 2000).

The notion that sustainability and food security can be achieved within a capitalist economic structure runs up against two issues central to alternative agrifood movements: environmental sustainability and social justice. The idea that either nature or people can be sustained within this economic system disregards the expansionist logic of capitalist dynamics. The foundation of capitalism is an imperative for economic growth and the substitution of less profitable for more profitable means of production—irrespective of either environmental or social consequences (Atkinson 1992). As Smith (1984: 268) has bluntly stated, the motor of the capitalist mode of production is, for the first time in history, "accumulation for accumulation's sake." In capitalist economies, profits are both the means and end of economic activity. Profit comes from expansion of production, new markets, and reducing costs.

In agriculture, as in other industries, growth and expansion are central, requiring agricultural enterprises to seek new sources of profits and new markets (de Janvry and LeVeen 1986). The same search for profit applies to farming practices and inputs. As with conventional agriculture, technological advances in organic farming tend to be those which rely less on cultural practices and more on commodities—things that can be sold at a profit. Organic farming publications are filled with advertisements for expensive biological inputs such as pesticides, compost, beneficial insects, and soil amendments (Rosset and Altieri 1997). Within the existing economic system, commodifiable solutions—those which can be profitably manufactured and sold—are the solutions to agricultural problems that get developed, marketed, and used (Bird 1988).

Is it possible to achieve environmental sustainability within a capitalist economic structure? This is the position of many in alternative agrifood movements. According to O'Connor (1993), reformist Greens posit that capitalism can be reshaped so that it is consistent with the sustainability of nature. Others are not so sure. For Benton (1989) the internal contradictions of capitalism are made most concrete in production processes such as agriculture that are "ecoregulatory." Benton characterizes ecoregulatory practices as those for which (1) the subject of labor is the conditions for the growth and development of the product, not the transformation of a raw material into a product; (2) labor is applied to regulating and reproducing, rather than transforming raw materials; (3) the time and space characteristics of the labor activity are shaped by organic developmental processes; and (4) nature-given conditions are both the conditions of the labor process and subjects of labor. In this framework, as "nature" becomes commodified under advanced capitalism, problems of sustainability deepen. These conditions occur when efforts to defend or restore profits ignore the maintenance of the environmental conditions of production. These conditions of production include soil and water—which cannot be produced or reproduced capitalistically, yet are bought, sold, and used as if they were commodities. In the end, efforts to cut or externalize costs reduce the long-term productivity of the environmental conditions of production, raising long-term average costs. For example, the introduction of cost-cutting pesticides in agriculture has produced pest resistance to pesticides, which has in turn created additional costs.

Capitalism runs up against environmental limits only in the form of economic crises; natural limits are simply a barrier to accumulation that must be overcome (O'Connor 1998). There is no profit in conservation for an individual firm. Even organic farmers who have a strong commitment to environmental conservation or food safety are trapped in an economic system in which they must purchase inputs and sell their products. As Bird (1988) pointed out more than a decade ago, once market competition among organic farmers develops, organic farmers will be driven by economic efficiency rather than environmental rationality. It is unclear how this incompatibility between profit and the environment can be overcome within the logic of a "free market" economy.

The same problem applies to social justice. Aside from the dynamics of production and organic foods marketing, a "free-market" approach is antithetical to achieving food security. O'Neill (1986: 107) points out, "If there are unrestricted economic rights to run life on commercial lines and accumulate private property, there cannot be rights to food or welfare." In a truly free market system no state actions should be taken to mitigate the hunger of those who lack the means to demonstrate their "effective demand" for food, that is, the ability to pay for it. They simply go hungry, since they serve no purpose in an economic system that regards food as a commodity (George 1985).

In the age of neoliberalism, hunger has again become a matter for charity rather than for state action, and the issue of hunger in wealthy Western societies has been depoliticized (Riches 1997). Charity was the approach taken for hunger relief before federal food assistance programs were introduced. At that time food assistance consisted of voluntary organizations serving neighborhoods or specific religious or ethnic groups (Poppendieck 1986). Interestingly, the Secretary of Agriculture Dan Glickman's 1996 announcement of the Community Food Project grants stated, "These grants will enable 13 communities to implement their own ideas for helping their neighbors." This statement sounds like a long step backward in history. Community food projects are primarily focused on developing strategies for reducing dependence and increasing self-reliance. Less attention is given to understanding and changing the system that has produced food insecurity in the first place. So far, community food projects have avoided addressing basic social contradictions or inequities. Instead, they tend to embrace concepts of decentralization and self-reliance, seemingly reflecting an affinity with contemporary individualistic, neoliberal approaches to solving social problems.

Of course, meeting social needs is neither the goal nor the function of a capitalist economy. Maximizing profit requires the subordination of use value such as nutrition to exchange value (money). A clear example in the agrifood system is the production of "junk" foods that have incidental or negative nutritional qualities and health effects, but very high profit margins. In many impoverished countries, agricultural production has been shifted from meeting basic food needs to producing crops for profit, reducing local food production and consumption (Kirkby et al. 1995). In Mexico, for example, increases in the production of livestock feed for the global market has displaced the production of basic foods and worsened conditions for the already malnourished. Far from serving people's food needs, growth in agricultural productivity has resulted in "continued immiseration and the creation of scarcity" (De Walt 1985: 54). Another example is diet foods, one of the fastest-growing segments of the American food industry. Americans spend $5 billion every year on special diets to lower their calorie consumption, while 400 million people worldwide suffer physical and mental deterioration from undernourishment (Durning 1990). Diet foods are a paradoxical source of profits in a world rampant with hunger (Friedmann 1995).

In a capitalist economic system, natural resources, labor, capital, technology, and food have all become commodities that are sold and bought at a price set by the "free" market. The insatiable search for profit has created negative environmental and social consequences. As Karl Polanyi observed long ago in *The Great Transformation* (1944), "While production could theoretically be organized this way, the commodity fiction disregarded the fact that leaving the fate of the soil and people to the market would be tantamount to annihilating them." Achieving agricultural sustainability and food security requires changing the social relations and material processes that structure and maintain the conditions of production and people's access to resources.

Ideologies of Class and Merit

If there is one concept that neoclassical economics refuses to address, it is that of social class, particularly as defined as relationship to the means of production such as land ownership. Alternative agrifood movements have also tended to turn away from the issue of class, in that they rarely address the material interests or forces behind the ideology of economic liberalism. In these movements, economic critique of the conventional agrifood system generally revolves around issues of corporatization, globalization, and industrialization. In our California study of alternative agrifood institutions, for example, interviewees tended to see problems and solutions in the agrifood system as centered much more around these kinds of economic issues than around class or the fundamental dynamics of the market economy.

Many alternative agrifood advocates see corporations and industrial agriculture as key obstacles to the achievement of sustainability and community food security. Henderson (1998), for example, decries the restructuring of world food system under corporate control. McMichael (2000) speaks in terms of the "excesses" of the industrialization of the food system, and Clancy (1997) refers to the "invisible costs" of industrialized farming. When health issues such as nutrition and food safety are mentioned, these are often linked with the industrialization and mass production of food (e.g., Gilg and Battershill 1998). Murdoch and Miele (1999) link environmental problems in agriculture to the use of industrial agricultural techniques.[3] In alternative agrifood movements the corporate organization of production and land ownership is seen as problematic, as is the industrialization of the production process.

Other inequitable economic relations, such as the relationships between farmers and brokers or those between family-farm owners and hired farm-workers, are obscured or rarely addressed. Corporate farms are often portrayed as villains while family farms are often depicted as the ideal. Yet both family farms and corporate farms share an economic position in the sense that they are generally entrepreneurial, property-owning firms that hire workers. Farmers, whether family farmers or corporate farmers, are a distinct class from farm laborers, who have only their own labor power to sell. This is the fundamental distinction of class and relationship to the means of production.

There is little in alternative agrifood movement discourse that suggests a critique of private property as a fundamental economic relation or seeks redress for historically inequitable land acquisition patterns. The economic orientations of alternative agrifood movements focus more on maintaining the conditions of agricultural property and production for those who currently possess them (e.g., family farmers) than on improving conditions for those in less privileged economic positions. As discussed, farmers are the focus of the sustainable agriculture movement. And farmers tend to be private property owners whose political positions are "seldom consistent with reducing the prerogatives of property" (Magdoff et al. 1998: 12). Alternative

---

3. A similar pattern has been observed in the environmental movement. According to Atkinson (1992), a key difference between environmentalists and those more concerned with social justice issues is the extent to which they see industrialization or capitalism as the primary root of environmental problems.

agrifood advocates tend to view the social relations of private property unproblematically. For example, community food security activists have noted the increasing abandonment of inner-city land as cities expand through outward development, yet the conditions of "ownership" of this land is rarely questioned. Although such movements exist in Europe, in the United States, there are few efforts for "squatters' rights" or transforming abandoned land into a food-producing "commons."

A crucial component of the ideologies of individualism and economic liberalism is a naturalization of social relations of ownership and hired labor. Yet who is able to own property and who must hire out their labor have not been decided by merit, but rather by accidents of history and inequitable public policies. For Williams (1973: 7) the "laws, constitutions, ideologies, which are claimed as natural, or as having universal validity or significance, simply have to be seen as expressing and ratifying the domination of a particular class." There is nothing natural or necessarily "earned" about present patterns of land ownership. These patterns were initiated and continue to be enforced by the state through a legal system of private property ownership. The precedent established by the enclosure of common lands in Europe (dating from the 1200s) was reproduced in early expansionist U.S. policies. Obviously this expansion depended upon seizing land from native inhabitants, usually by force or deceit.

Distribution of these lands favored the wealthy. In 1785 Congress passed the Basic Land Ordinance, which prescribed how land was to be surveyed and sold to farmers, and later passed a law encouraging the expansion of agriculture into "frontier" areas. This ordinance of 1785 provided for the sale of federal land at auctions in minimum lots of 640 acres for cash only. This provision far exceeded the finances of a typical settler with the result that purchases were usually made by land speculators (Cochrane 1979).[4] Ability to purchase land became more skewed with the passage of the Land Act of 1796. This act maintained the minimum lot size of 640 acres, but doubled the purchase price. This helped to established a pattern of con-

---

4. It was not until the Land Act of 1820 that there was any attempt to provide land to "average" pioneers. This act lowered the minimum bidding price for land and reduced the minimum purchase size to 80 acres, although it also abolished credit for land purchase. Forty years later, the Homestead Act of 1862 provided settlers with free land provided they lived on the land and farmed it for five years. The Homestead Act helped to establish a pattern of owner-occupied family farms in the Midwest and West. Homestead units were originally 160 acres, but maximum acreage was later raised to create more economically viable units in areas where rainfall was uncertain.

centration in agricultural land ownership that continues to this day. Although American agriculture basks in the myth of having been settled by a cadre of small-holding owner-operators, ownership of productive land is actually highly concentrated. Only 4 percent of landowners hold 47 percent of American farmland (Census Bureau 1987), a level of concentration that rivals the most skewed in other countries. People who are poorly paid and who have not inherited property have virtually no chance of entering the ranks of landowners. Yet in the alternative agrifood movements, there is little discussion of how historically marginalized people can gain access to factors of production such as property and capital.

Not only is this inequitable, present property relations may work against reformist alternative agrifood efforts, such as community gardens. These gardens can be ephemeral, since they are built either on land owned by the municipality or a private entity that is not using the land at the time. Such land can be reclaimed at any time and on short notice. Changes in land-holding patterns are necessary for community gardens to become stable sources of food production. At a minimum, municipalities could make commitments to reserve land for urban agriculture. Often, however, the trend is in the other direction; many cities are opposed to the informal use of public lands for agriculture or need the economic return from land sales or leases to augment city budgets. For example, caught in a fiscal squeeze brought on in part by reductions in local revenues, the city of Santa Cruz, California, decided it needed to sell the city land occupied by the highly acclaimed Homeless Garden Project. The only way to ensure a community garden's survival is for it to either become a land trust (i.e., permanent open space) or receive permanent site status with the protection of the municipal parks department (Baker 1997). The structure of land ownership can also work against the poor in cities by enabling land uses that run counter to their needs. For many of the poor people who live in East Palo Alto, California, for example, leaving land abandoned may be preferable to development options. The building of the city's first shopping center led some landlords (speculating on future development) to raise rents dramatically; this led to an increase in hunger as low-income tenants then had less money available for food (Richmond 1998).

The privatization of land in America created a class of owners and a class of workers. Those who did not acquire land or who lost their land became wage laborers. Today farmworkers are paid low wages, suffer difficult working conditions, experience hunger, and live in substandard housing; many are vulnerable in their employment and citizenship status. In some agricultural regions one might ask how different today's farm labor conditions are from those described by Mintz (1995: 12) on an eighteenth-century sugar plantation: "The conditions under which landless people worked were determined by others: the hours when they might eat or rest, where they took their food, how they got to and from work." Ironically, a 1997 USDA pamphlet on sustainable agriculture extols George Washington and Thomas Jefferson as model sustainable farmers who searched for alternative methods to improve "the lives and health of the citizenry" even though both owned many slaves who worked their owner's fields. While the idea of a small-holder agrarian democracy is attributed to Jefferson, he himself did not represent this ideal. He continued to expand his plantation throughout his life, and at his death it covered ten thousand acres, which were worked by 150 slaves (Esbjornson 1992). As Danbom (1997) points out, the majority of people—women, including slaves, farm laborers, indentured servants, and tenants—were invisible in the agrarian world Jefferson promoted, since the category of "farmer" was based on property ownership.

Then, as now, many of those living in the worst poverty, holding the most dangerous jobs, are farmworkers. Yet a large proportion of American farmers have traditionally been hostile to improved conditions for farmworkers. For instance, agrarian capital in California has spent more energy and organizational effort on managing the labor supply than on any other aspect of the production regime (Deshpande 1991). Growers have been able to exploit outcast, "unassimilable" races who came to California throughout the second half of the nineteenth and first quarter of the twentieth centuries as inexpensive farm labor. The earlier history of the importation and exploitation of agricultural labor in California is told in the classic *Factories in the Fields* (McWilliams 1939). Wages were kept low in part by the discursive practice of racism (Deshpande 1991), and continue to be kept low by the orchestrated oversupply of labor. U.S. policies also engineered an abundant, docile agricultural labor force by encouraging the immigration of foreign laborers and developing guest worker programs (e.g., the *bracero* program). Under the *bracero* program the federal government paid for the workers' transportation to and from the United States and paid their medical, unemployment, and disability expenses (Jelinek 1982). In this way, the agricultural industry, primarily large farmers, received a labor subsidy in the form of cheaper immigrant labor not available to any other American industry. Most of the beneficiaries were large-scale farmers of California. In 1960 over 80 percent of the *braceros* worked only 5 percent of California farms.

Many in alternative agrifood movements seem to accept this structure of hired farm labor unproblematically. For example, the Asilomar Declaration for Sustainable Agriculture states: "Healthy rural communities are attractive and equitable for farmers, farmworkers, and their families" (Committee for Sustainable Agriculture 1990). This statement assumes the necessity of present social relations in the production process and does not recognize that farmers and farmworkers often have different interests because of their different economic and social positions. The statement also reflects a high degree of comfort with an economy based upon farm owners who hire landless laborers to plant, tend, and harvest crops. In California, at least, nearly all farmers hire farmworkers, including csa growers and former farmworkers who have become farmers.

Where farmworkers are mentioned in the discourse of sustainable agriculture, they are often objectified and treated primarily as economic inputs, along with equipment and fuel. For example, the same alternative agriculture studies that closely detail natural phenomena such as plant/insect interactions tend to ignore human/human interactions, treating farmworkers as just another cost of production (e.g., nrc 1989). There is little or no discussion of who the workers are, their working conditions, or their wages. At times, the antagonistic and exploitative approach to labor has carried into alternative agrifood movements. For example, the California Certified Organic Farmers organization was instrumental in defeating a farmworker labor bill supported by labor rights organizations (Buck et al. 1997). The bill would have expanded the ban of the short-handled hoe to include a ban on working with bare hands. While the adoption of sustainable agriculture practices may reduce worker exposure to toxic chemicals,[5] until recently, there has been little effort to deal with farmworker issues such as such as low wages or poor housing.

This situation contrasts sharply with the priorities of some alternative agrifood organizations in California during the 1970s. At that time the

---

5. While this claim is mostly accurate, some efforts to make agriculture more sustainable for some people can cause greater harm to workers' health. For example, Wright (1990) showed that consumers' concerns about the health effects of pesticide residues prompted growers to shift to chemicals that do not persist in the environment, but which can pose a greater immediate threat to the health of farmworkers and people living close to fields. Some of the compounds allowed in organic agriculture can also adversely affect human health. For example, sulfur, used extensively in grape growing is a relatively low-toxicity substance allowed in organic farming. However, it can irritate skin, eyes, and lungs and causes illness. In California, workers in grape production suffered one of the highest numbers of pesticide-related illnesses (Reeves et al. 2002).

search for alternatives included alternatives to not only environmental destruction, but also the poverty and racism inherent in the agrifood system (Allen et al. 2003). In their more recent incarnations some of these organizations have modified their approaches to labor issues. For example, the Agrarian Action Project became less active in support of farmworkers when it merged with the California Association of Family Farms in 1993 to create the Community Alliance with Family Farmers. Once dedicated to farmworker advocacy, the organization's emphasis changed to emphasize farmer-to-farmer education, biological strategies for pest and fertility management in cropping systems, and direct marketing. An interest in farmworker issues may be resurfacing, however. For example, the California Sustainable Agriculture Working Group's board of directors includes a representative from the UFW in an effort to include the perspectives of farmworkers in sustainable agriculture priorities and planning.

Can an environmentally sustainable agriculture be developed by overworked and underpaid farmworkers? Possibly. Thompson (1995) observes that the slave agricultures of Egypt and Sumeria were likely quite sustainable, judging by production and resource, rather than social criteria. Alternative agriculture movements will need to decide if they want to elevate environmental goals over those of social equity. At present, environmental priorities tend to dominate in the movements.

The Fetishization of the Environment

What distinguishes the sustainable agriculture movement from early forms of agrarian idealism is its prioritization of environmental problems (Esbjornson 1992). According to Buttel (1993a), the sustainable agriculture movement probably would not exist without the growth in environmentalist sentiment and the environmental movement that arose during the late 1960s and early 1970s. Discourses of alternative agriculture are infused with environmentalist perspectives and approaches, particularly the privileging and essentializing of nature. Hamlin (1991: 508–9) finds the appeal to nature to be a "peculiar feature" of the consideration of alternative agricultures. For example, one of sustainable agriculture's earliest and most influential proponents and researchers, Wes Jackson, bases his approach to sustainable agriculture on "nature as analogy" (1990). In this conception of nature, the environment is considered to be a physical space and set of laws that exist and operate external to and independent of humans. Crews and

others (1991: 146) state that although the profitability of sustainable agricultural systems is constrained by the social structure of agriculture, "sustainability itself is constrained solely by the ecological conditions of
agriculture." These approaches prescribe saving nature from the negative
effects of human actions. Most alternative agrifood advocates, however, call
not for a hands-off approach to nature, but for a more benign configura-
tion of nature-manipulating strategies. Lehman and others (1993: 139), for
example, define sustainable agriculture as consisting of "agricultural
processes that do not exhaust any irreplaceable resources which are essential to agriculture," with agricultural processes being "processes involving
biological activities of growth or reproduction intended to produce crops."

Others go further, suggesting that ecological principles explain human
behavior or should be used to design society. Somma (1993: 372), for example, writes that concepts such as system stability, carrying capacity, and niche
"present insights into human behavioral requirements in production and
reproduction." Both "deep ecology" and bioregionalism are philosophies
prominent in alternative agriculture. Deep ecology espouses a reverence
for the natural world and is based on holistic principles for managing
human intervention in the environment. Bioregionalism has a similar orientation, focused on a geographic scale. A bioregion is "a life-territory, a
place defined by its life forms, its topography and biota, rather than by
human dictates; a region governed by nature not legislature" (Sale 1985).
Bioregionalists advocate developing a culture and economy based on the
ecological characteristics of a given region. According to Frenkel (1994:
289), "Bioregionalism is the belief that social relations ought to be derived
and governed by the local biophysical environment." Ironically, protecting
nature from society's control and management is considered integral to preserving the essence of human nature.

These approaches fetishize the environment rather than seeing it as a
human construct. In the application of ecological theory to human society,
"ecology becomes the foundation of and restriction of political possibilities, determining which values should guide politics and the forms politics
should assume" (Hayward 1994). Extrapolating human values from those
we perceive in nature can have regressive effects. Certainly, the idea that
biology is somehow destiny is antithetical to the perspective that categories
such as race and gender are socially constructed. Ecological feminism, for
example, which contains many strains and theoretical perspectives, is unified in its belief that the root of the degraded condition of both women and
nature is men's efforts to dominate both. Feminist political ecology chal-

lenges the gendered discourse of environmental science and, in some formulations, privileges the particular knowledge about the environment that women may possess, not because of their biology, but as a consequence of their gendered roles and experiences.

What is natural about nature? "Nature" itself is a social category and construct. Through their labor, humans exist at the center of a nature that they produce as they produce society. As Smith (1984: 18) points out, "The relation with nature is an historical product, and even to posit nature as external to society (a primary methodological axiom of positivist 'science,' for example) is literally absurd since the very act of positing nature requires entering a certain relation *with* nature." Thus, the environment is less a physical "fact" than it is a set of humanly mediated relationships (Redclift 1987). Human knowledge of nature can only ever be a *relation* between people and nature. The idea of nature accepted in a society at any given time is always a reflection of human relationships prevailing in that society during a particular period (Haila and Levins 1992).

While there is a "nature" that preexisted human beings, and people are subject to natural forces such as gravity and time, the nature that we talk about in relation to agriculture is a humanly reconstructed nature. Agriculture is an intentional, human productive activity that has always been socially organized and becomes more so as it develops. Smith (1984) shows that humans produce their means of existence from nature and at the same time generate additional needs that require further production, which in turn leads to further divisions of social labor. People's relations with nature then become mediated through the social institutions designed to regulate production and the distribution of surplus. As Friedmann (1993: 213) puts it, "From the first domestication of plants and animals, humans irreversibly posed for themselves the problem of creating social relations through which to act in concert upon nature."

Nature as we know it is thus a dialectical process of transformation between humans and their environment—nature is "produced" through human labor as well as self-producing. People apply human labor and appropriate from nature in order to transform natural materials into forms that are useful to humans. Elements of the biosphere (e.g., soil, water, and energy) become "resources" only when people define, use, and exchange them as such. Redclift (1993) emphasizes that human concerns about the sustainability of the resource base make sense only in relation to the human agents who manage the environment. A clear example of this is how animals are considered in alternative agriculture. While the sustainable agriculture movement has sometimes raised issues about the humane treatment of livestock, it has not questioned the ethics of eating animals in the first place (Regan 1993). A distinction is made between "wild" animals, which are to be preserved, and "domestic" animals, which are raised precisely to be destroyed and sold as commodities. The latter is somehow not regarded as a rationalist degradation of nature. Interestingly, while "first" (pristine) nature is to be preserved, "second" (mediated) nature is afforded no such protection.

Critics of ecological essentialism argue that biology can neither explain relations between people nor be used to derive principles for politics. According to Haila and Levins (1992), ecology tends toward naturalism and objectivism and to remain outside any historical framework. Using food as a heuristic, Martinez-Alier (1995) illustrates the need to "historicize ecology." He points out that while biology determines the number of calories required for human survival, what and how much a person actually consumes is determined not by "nature" but by politics, economics, and culture, all of which contribute to the large differences between rich and poor. General attributions of ecological problems to humans as a species "invite us to overlook oppressions and divisions within the human community and are ethically irresponsible if they imply that the cause of nature should be promoted at the cost of a concern with social justice and equity in the distribution of resources" (Soper 1995: 13). Deterministic ecological models not only fail to historicize the distribution of resources, they can also obscure the possibility of changing distribution in the future. These kinds of ideological orientations have tended to produce "often vague environmentalist sympathy that is inadequate to political questions about power and justice" (Darnovsky 1992: 50).

None of this is to deny that nature and environment are central to human existence and the agrifood system. All food production requires that nature be utilized by humans, and no one in alternative agrifood movements opposes the use of nature for human ends. While agriculture obviously depends upon nature and natural processes, it is an inescapable fact that laws, watersheds, political parties, and bioregions are all decidedly human constructs. Of course the material world *does* exist. Watts and Peet (1996) caution against going too far down the path of the social constructionist notion of nature because it tends to overestimate the power humans have to transform or manipulate nature.

The relevant distinction in the agrifood system is not so much between what is natural and what is social, but rather between what can and what cannot be reconfigured and improved upon. Which priorities we choose to pursue in food security and sustainable agriculture can only be posed as social questions. They cannot be decided through appeals to nature but only though the political process. Bryant (1991: 164), in referring to environmental managerialism, points out that the central issue is who formulates and implements environmental strategies, and in whose interest are these strategies? If environmental problems are seen as humanly constituted and historical—and we deny that there is a "natural" basis for the current social order—such problems can only be seen as a the result of a historical process that is the accretion of human decisions, political processes, and chosen distributions of material and cultural goods. Thus, different decisions, different political processes, and different distributions of resources are not only possible, but necessary, for solving them.

Political ecology may work as a new epistemological approach for alternative agrifood movements and institutions. The aim of political ecology is to uncover the root causes of environmental and human resource degradation by studying the interaction between human society and nature. Fundamental to this method is the assumption that these root causes are not simply biological or technical and that the problems cannot be solved by technological fixes. Because of its multifaceted subject matter, political ecology is necessarily interdisciplinary. As its name implies, it combines the methods of ecology and political economy. Its methodological holism and its attention to natural systems come from ecology. However, ecological methods alone are insufficient because humans cannot be studied in the same way as other species. Human actions are shaped by politics, power, ethics, culture, and so on, all of which would be missed by a simplistic application of ecological concepts to society. Because these aspects of human societies have long been the subject of political economy, political ecology draws some of its methods and concepts from that tradition.

Political ecology avoids fixed, deterministic models, analyzing problems in historical context, which usually makes it clear that "systems" change and are changeable. Strict systems analysis might help us to see the effects of changes in inputs to the system, but it discourages us from imagining changes to the system itself. Only when analysis avoids deterministic models does it help us imagine fundamental change. Political ecology employs two basic levels of analysis (Thrupp 1993). At the local level the research begins with an analysis of the relations between society and nature. For this analysis, the researcher often uses anthropological or sociological methods such as interviews or ethnography. At the structural level political ecology aims to understand the larger historical and socioeconomic forces in the wider political economy that underlie and situate the specific causes of the local problem. Moving toward an environmentally sound and socially just agrifood system will be predicated upon discovering causes behind the symptoms. If we do not look at causes, we are likely to repeat effects.

Alternative agrifood movements include residual, dominant, and emergent ideologies. Residual and dominant formulations such as agrarianism, individualism, economic liberalism, and class and merit persist within alternative agrifood movements. This has an advantage in that alternative agrifood movements have been most tolerated where they present little challenge to the institutional and ideological formulations of the dominant culture. However, these ideologies tend to reinscribe existing social relations that are counter to increasing equity and broadening access to resources and power. Ideologies that support the status quo or blame people for creating their own problems reduce the impetus for progressive change. These become embodied in everyday forms of discourse that then express and normalize existing power relationships. The ecological fetishism inherited from the environmental movement can also retard progress toward the social justice goals of the movement by deflecting attention from social issues.

The point, though, is that ideologies are just that—ideas. By replacing the idea of "values" with that of "processes of valuation" we can begin to understand how these processes operate, and better understand how and why certain kinds of "permanence" get constructed in particular places and times such that they form dominant social values to which most people willingly subscribe (Harvey 1996: 11). One of the key issues with the processes of valuation in alternative agrifood movements is the relatively narrow range of those who participate, and therefore whose ideas and priorities are represented. Not all members of the agrifood community, namely, farmworkers, food service workers, and consumers are proportionately represented. Even for those who are represented, not all voices and perspectives are considered equally. Developing more inclusive discourses and practices of sustainability and sustenance will require a deeper democratization of the movements. It also requires examining the practice of democracy and role of participation in alternative agrifood movements.

6

participation and power in alternative agrifood movements and institutions

Alternative agrifood movements and institutions have long been committed to the philosophy and practice of democratic process. This is seen simultaneously as a goal of environmentally sound and socially just agrifood systems and as a method for developing these systems. By democratic process I mean several things: inclusion of those who are involved in a situation or would be affected by decisions made, voting equality, opportunity to develop priorities, and equality of access to information. Alternative agrifood movements and institutions engage all of these aspects of democratic practice as central principles.

This chapter focuses on the concept and practice of democracy in the context of alternative agrifood movements and institutions, first highlighting examples of democratic practice within alternative agrifood efforts. This leads to the consideration of areas with particular implications for the democracy of alternative agrifood movements and institutions. I begin by exploring the extent to which various categories of voices are considered legitimate and audible in alternative agrifood efforts. Next I examine the issue of gender because of agriculture's historical simultaneous dependence upon and exclusion of women. I then turn to questions about the relationships among democracy, power, and privilege. I explore another area relevant to discussions of democracy in alternative agrifood movements—food system localization—in Chapter 7.

Democratic Practice in Alternative Agrifood Movements and Institutions

Commitment to the ideal of participatory democracy can be observed in many places within alternative agrifood movements and institutions. These range from the inclusion of a broad range of people in national-level planning and programming to the engagement of those in local communities to learn about and change the agrifood system.

In alternative agrifood programs established within traditional institutions, the inclusion of diverse constituencies has been a high priority. The USDA SARE program, for example, strives to operate on "principles of inclusion, partnership, and participation" (SARE 1998b: 3). The SARE Western Region group clearly states that "sustainable agriculture research and education involves a battery of agricultural and environmental scientists; the 'in-the-field' experts themselves, farmers and ranchers; as well as economic and social experts; and the end customers, the public" (Western Region SARE 1995). Apparently this degree of diversity is a departure from how things have traditionally been done in the USDA. SARE asserts that this "broad representation remains largely unique in federal grant funding for agriculture" (SARE 1998a).

An additional way that SARE works toward democratic process is through decentralized decision making. SARE operates through four regional committees, and the membership on these committees spans a wide range of agrifood system participants. The administrative councils and technical review committees in SARE's four regions include farmers, ranchers, and agribusiness; public and private research and extension institutions; nonprofit organizations; and government agencies. Not only does this provide a forum for multiple voices, it also fosters discussion and cooperation among groups of people who otherwise might not have come together. A U.S. General Accounting Office (1992b: 35) review of the SARE program found that "many people we spoke to who were involved in or knowledgeable of SARE said that the most dramatic benefit of the program was the opportunity for these often opposing groups to meet and work together on setting priorities and approving proposals." This was a new experience for some of the committee members. One SARE participant admitted that at first they found the level of diversity on the committee frightening; ultimately, however, they came to believe that this diversity provided "strength, energy and creativity" that the committee might otherwise have lacked (SARE 1998b: 1). Producers and nonprofit representatives have been selected to serve as chairs of SARE administrative councils, showing increasing levels of trust and respect among the participants since the beginning of the program (Dyer 1999).

UC SAREP and Kellogg also promote the importance of inclusiveness and diversity in their programs. UC SAREP points out that all agrifood system participants—farmers, farm laborers, policymakers, researchers, retailers, and consumers—have particular and important roles to play in developing a sustainable food and agriculture system. This commitment is put into practice in a number of ways. For example, while at first UC SAREP's Public Advisory Committee was heavily weighted toward farm interests, it has become much more diverse over time. It now includes a large proportion of public interest organizations, including environmental, consumer, and rural issues groups along with farmers and farm organizations. The Kellogg Foundation places a similar emphasis on diversity and inclusiveness. For example, a primary goal of the Kellogg IFS program was to ensure the representation of "traditionally underserved people, such as women and minorities," whose resources are often more limited and who may as a result be "less able to take a chance on new farming practices." And while Phase 1 of the IFS program focused primarily on farming, the Phase 2 projects were much broader in scope.

The alternative agrifood policy organization, the National Campaign for Sustainable Agriculture, actively sought to diversify the people and interests involved in the formation of the 1995 farm bill and continues to do so. The Campaign's approach to sustainability includes social as well as environmental goals, and it actively works to include those who have previously had little or no voice in national agricultural policy. Women and ethnic minorities play significant roles in the Campaign, which is composed of hundreds of grassroots and national organizations and includes representation of family farmers, environmentalists, farmworkers, consumers, and animal protection advocates (National Sustainable Agriculture Coordinating Council n.d.). In all of these cases—SARE, UC SAREP, the IFS projects, alternative agrifood institutions, and SAWG—people who otherwise might not have crossed paths worked together and found the experience to be broadening and rewarding.

Democratic principles are also being translated into practice in community food projects. For example, the community-based food policy councils and similar efforts to integrate food and farm policy taking shape throughout North America bring together diverse constituencies to work on improving agrifood policies. Many begin with the assumption that nutritious food is a basic right of all citizens and advocate that governments and community groups take active roles in planning for and ensuring food security (Ashman et al. 1993).

We found in our study of alternative agrifood institutions in California that they also tended to be both diverse and ecumenical. Taken as a group, these organizations engaged a broad spectrum of the population, including affluent and low-income consumers, urban and rural people, farmers, workers, businesspeople, and students. This range of constituencies was often found within individual organizations as well. In fact, the leaders of several organizations marveled at the degree of class, cultural, ethnic, and religious cooperation that has emerged out of their food-based projects.

In addition to including a chorus of voices and providing frameworks for cooperation, many alternative agrifood projects increase people's knowledge about the agrifood system. For example, through community food assessments—a process developed by the community food security movement—people can develop much more comprehensive and also personal understandings of how the agrifood system works. A community food assessment is "a collaborative and participatory process that systematically examines a broad range of community food issues and assets, so as to inform change actions to make the community more food secure" (Pothukuchi et al. 2002: 11). Such assessments provide opportunities for people to understand their place in the agrifood system, as well as the factors that constrain or enable their access to resources in the system.

Another approach that has gained currency is the "foodshed" concept. A foodshed is a locally based, self-reliant food system that works with rather than against the ecology of the region. For Kloppenburg and others (1996) one of the primary powers of the foodshed concept is that it can provide a connection between theory and action. Foodshed analysis includes examining the structure and dynamics of the global agrifood system, identifying and studying emergent alternatives to it, and working to link these elements. This kind of approach can be profoundly educational and empowering and have great prescriptive value. The knowledge gained through community food assessments and foodshed analyses can, in turn, catalyze further civic participation. Food-system issues that may have seemed abstract and remote become more real and more personal. Until problems are perceived as more than abstractions, they are unlikely to inspire committed efforts to solve them.

Community actors and students using food assessment and foodshed approaches cannot help but see inequities in resource control and access embedded in their local food systems, an understanding which can help empower them to work to change these inequities. Even projects that allow responsibility for food security to devolve from the federal to the local level can contain within them the potential for people to participate in actively reinventing rather than passively accepting the food system. Moving in the other direction, although food policy councils have functioned primarily at local or regional levels, their priorities and organizational strategies also pertain to democratic decision making at national levels. Projects that include community organizing and food councils can increase self-determination in food issues, a process of politicization that builds networks with the potential to engage other areas of civic life and political issues. This is a deep kind of democratization.

Audible Voices in Alternative Agrifood Movements and Institutions

It is self-evident that decision-making groups with a narrow range of participants can address only a narrow range of issues and options. Individual perspectives, which arise out of differences in socioeconomic background and day-to-day experiences, play a pivotal role in what decision makers see as problems and solutions. People cannot have a voice if they are not included in the discussion. In this regard, alternative agrifood movements are in the position of needing to rectify imbalances in the demographic profile of who has participated in the traditional agrifood system. In addition, building a democratic movement will involve attending to issues that concern the full range of participants in the agrifood system as a corrective to the traditional emphasis on farmers over other agrifood system participants.

On the issue of demographics, it has been observed that alternative agrifood movements—particularly sustainable agriculture—tend to be similar to mainstream environmental movements. That is, they are disproportionately European-American and affluent. Agrarian-based movements and institutions have huge historical barriers to overcome in this regard. For example, government agencies related to agriculture have had extremely homogenous gender and racial compositions. No other federal agency ranked lower in hiring and promoting minorities than the USDA (*Kansas City Star* 1991). In 1992, 89 percent of senior-level USDA employees were white, and 82 percent were male (U.S. Office of Personnel Management 1992). The percentage of European-American males was even higher in senior executive positions. The gender and ethnicity inequities were significant enough that the USDA created a Civil Rights office to "facilitate the fair and equitable treatment of USDA customers and employees while ensuring the delivery and enforcement of civil rights programs and activities."

Other key decision makers in agrifood institutions are the researchers who decide which problems are worthy of study and which are not. As with other agrifood institutions, agricultural research has been dominated by European-American men. In 1976, 99.6 percent of agricultural scientists were male, and the agricultural sciences had the highest percentage of whites—98.6 percent—of all scientific fields in the United States (Busch and Lacy 1983). Ten years later, while the proportion of ethnic minorities in the agricultural sciences had achieved parity with that of other sciences, women were still underrepresented. These kinds of imbalances clearly violate the premises of democracy.

The priority that alternative agrifood movements and institutions place on democracy is all the more remarkable given the historically undemocratic characteristics of American agrifood institutions. In other ways, though, the movements and institutions reproduce a long-standing privileging of the priorities of only one group of those who labor in the agrifood sector—farmers. This emphasis on farmers is out of proportion to their numbers among agrifood system workers. Of those who work in the agrifood sector, only 7 percent are farmers and farmworkers directly involved in agricultural production. The other 93 percent of agrifood system workers have jobs in other sectors of the system: transportation (3 percent), food processing (9 percent), equipment and inputs (19 percent), food service (35 percent), and food wholesaling and retailing (38 percent) (derived from Edmondson 2003). And in California at least, most of those who are involved in farming are farmworkers, not farm operators. In California there are eighteen farmworkers for each farmer, and hired farmworkers perform at least 80 percent of all the farm work in the state (Villarejo 1990). Even in Rodale's extraordinarily comprehensive study and set of recommendations for the U.S. agrifood system, *Empty Breadbasket?* (Cornucopia Project 1981), workers scarcely rated a mention. Its ten goals for the food system were abundance, dependability, sustainability, safety, efficiency, appropriateness, equitability, wealth, flexibility, and openness. Even with this degree of comprehensiveness, workers remained invisible. The only discussion of workers was a recommendation that "the food system should not endanger workers, consumers or the environment" (Cornucopia Project 1981). Workers are simultaneously everywhere and nowhere in the agrifood system.

The emphasis on farmers over other workers in the agrifood sector is common in alternative agrifood movements and institutions. For example, the National Research Council's (1989) report, *Alternative Agriculture*, suggests that farmers reduce, not only the off-farm inputs that pose the greatest potential to harm the environment, but also those inputs that harm "the health of farmers or consumers." The farmworkers, who often endure the greatest exposure to agrichemicals and have extremely high rates of occupational disability, are not mentioned.[1] On its website SARE includes farm and food industry workers as participants in sustainable agriculture. Yet the elaboration of this inclusion ties back only to farmers, stating: "Tying producers—and their products—to the local community and educating consumers about sustainable agriculture can underscore a farm's vital role in the community, engendering good will toward agriculture in an increasingly suburban society" (SARE 1998a). Farmers continue as the privileged subject while farmworkers and food-industry workers disappear from consideration. In our California study of alternative agrifood institutions we found that organizations started after 1980 were less likely to address the problems of California's migrant farm labor force than those of the 1970s (Allen et al. 2003).

And while membership in SARE's regional councils is more diverse than that of similar USDA committees, it still has been weighted toward producers and scientists, who prioritize agricultural production. For example, in 1995, the SARE Western Region's eleven-member administrative council was composed of three producers, three private consultants (an environmental consultant, an agricultural real estate consultant, and an agronomist consultant), and five university or federal agency scientists (Western Region SARE 1995). There were no "social experts" or representatives of the public on the council. Dyer (1999) reports that environmental

---

1. Agriculture is one of the most dangerous industries for workers in the United States, with 23 deaths out of 100,000 agricultural workers in 1980–89 (Arcury 1998). Between 1984 and 1995, there were 324,524 deaths caused by fatal diseases of employees in farming and other agricultural occupations (National Institute of Occupational Safety and Health 2002). Health problems for agricultural workers include arthritis, cancer, respiratory disorders, and injuries from machinery. For those who work in the fields, pesticide exposure is substantial, causing poisonings, reproductive problems, and death. Although there is no national system for tracking pesticide poisonings, the Environmental Protection Agency estimates that each year hired farm workers suffer up to 300,000 acute illnesses and injuries from exposure to pesticides (GAO 1992a). The problem also affects children. A study of migrant children working on farms in New York state found that over 40 percent of farmworker children interviewed had worked in fields still wet with pesticides, and 40 percent had been sprayed while in the fields (ibid.).

groups have not been formally represented, either on SARE administrative councils or on review panels. In a U.S. General Accounting Office (1992b) study of the involvement of various types of groups on the SARE administrative councils, technical review panels or committees, and ten regional projects, reviewers found that of those participating, 26 percent were producers, 56 were scientists, and only 7 percent were representatives of nonprofit organizations (Table 6). This distribution became even more skewed toward producers and scientists at the level of regional project implementation. It is not possible to know if women and ethnic minorities were better represented on the committees than on traditional agricultural committees because no data were presented on participation by women or ethnic minorities.

The disproportionate emphasis on agricultural production is both a cause of and the result of some alternative agrifood priorities and premises. There is a tendency to privilege farmers as agents of change, the rightful beneficiaries of that change, and the savants who know what is to be done and how to do it. Most alternative agrifood advocates see farmers as the central figures. The importance of farmers and a family farm structure is assumed as a basic premise, rather than argued as a proposition. Roberts and Hollander (1997) argue that characteristics of the family farm particularly suit it to be the social basis for achieving agricultural sustainability. In their view, family farms can best adopt new sustainable technologies, which in turn will make them more successful.

The W. K. Kellogg Foundation's broad approach to agrifood system actors also seems to contract toward farmers in the articulation of IFS program goals. The first is to "help farmers integrate into their crop and livestock systems new methods that are productive, profitable, and environmentally sensitive. Those methods would also benefit the personal health of farmers and their families." The second is to "help people and their communities overcome any barriers that might otherwise prevent them from adopting improved, more sustainable agricultural systems." While lower priority was given to projects that focused solely on developing technologies than to those that also focused on the adoption of these technologies, farm technology remains at the core. Once again, the subjects and beneficiaries of sustainable agriculture are seen as farmers, and the path to achieving sustainability is through changing farm practices. Seventeen out of the eighteen projects explicitly mention producers, production practices, or farm families as the central focus in their statements of purpose. This constriction is not viewed as such by the Foundation, which requires that "the

Table 6 Types of groups participating in SARE

Groups	Administrative councils	Technical review panels	Regional projects	Total	Percent of total
Researchers/ extension personnel	19	53	99	171	56
Farmers/ranchers	7	23	49	79	26
Government personnel	18	14	1	33	11
Agents of nonprofit organizations	10	6	6	22	7
Total	54	96	155	305	100

Source: calculated from GAO 1992.

projects all involve a holistic approach to food production and distribution"
(W. K. Kellogg Foundation 1996).

Once again, this emphasis on farmers in the Kellogg IFS projects is
extolled by the outside evaluator for the projects: "These are farmer-driven
projects . . . universities and public policymakers are more responsive to
farmers than to paid advocates working for nonprofit organizations and
other institutions . . . extension agents and other professionals who aren't
farmers need to learn better how to 'be a guide on the side, not a sage on
the stage'" (Scheie 1997: 9).

The evaluator is concerned, in fact, that broader collaborations may
result in more conflict and less productive action. CASA recognized that the
original collaborating institutions were primarily oriented toward farmers
and farm practices. Since CASA also decided to concentrate on food security issues and farmworker issues, the group expanded its collaborators,
adding a nutrition program, California Adolescent Nutrition and Fitness
(CANFit), and a farmworker rights organization, Pueblo Unido, to the steer-
ing committee. The California project was the only one of the eighteen
IFS projects that did not explicitly focus on farmers or farm practices in its
statement of purpose. In other states, project statements specified improving the sustainability of farm practices, increasing the viability of family

farms, and linking farmers to other community groups, while little was said about how these community groups could benefit from these connections.

Similarly, SARE prioritizes agricultural producers as "agents of change" for sustainability as well as its subjects and beneficiaries. Its 2002 report states that "SARE works to increase knowledge about—and help farmers and ranchers adopt—practices that are profitable, environmentally sound and good for communities" (SARE 1998a). Farmers are explicitly the intended beneficiaries of SARE efforts; the program is "designed expressly to help farmers and ranchers find the answers they need" (SARE 1995). These farmers and ranchers, along with agricultural scientists, are viewed as the actors who will bring about sustainability. In the introduction to its ten-year report, SARE states that resources are "spent to help farmers and ranchers adopt [sustainable] practices" and that this has "made a difference in the lives of farmers and ranchers" (SARE 1998b: 3). No other agents, subjects, or beneficiaries are mentioned.

Farmers' interests tend to be privileged over those of other agrifood system participants even when this is not the intention. For example, despite their ecumenical language and inclusive statements about participation and democracy alternative agrifood movements often narrow to focus on farmers as the key agents and intended beneficiaries of agrifood system change. For instance, while Youngberg and others (1993) advocate the development of a vigorous sustainable agriculture coalition, the coalition they propose would be composed of "farmers and supportive groups." Another illustration of this phenomenon of the funneling toward farmers is a manual on promoting sustainable agriculture. The introduction to the manual states that developing a sustainable agriculture system requires that farmers, educators, activists, farm suppliers, and everyone else be included in a broad information exchange (Grieshop et al. 1996). The introduction further reviews a range of definitions of sustainable agriculture, most of which include concepts of social justice, humane working conditions, and economic equity. All of this shows diversity and inclusiveness in both topics and participants. Yet the remainder of the manual is devoted to cataloguing methods for developing "solutions that are consistent with growers' needs, concerns, preferences, and perceptions" and ways to link growers with farmer "innovators." No other audience is addressed, nor are any interests other than those of growers discussed. This focus on farmers and farming therefore excludes most of those working in the agrifood sector and privileges the voices of only a small subset of participants in the agrifood system.

As part of its efforts to develop ways that public research and education programs can become more equitable, transparent, and accountable, the Consortium for Sustainable Agriculture Research and Education (CSARE) undertook an evaluation of participation in USDA programs in sustainable agriculture. They found that while farmers were regular participants, there were "few signs" that consumers or environmentalists were represented as a group in the SARE programs (Dyer 1999). Farmer involvement in research projects in sustainable agriculture reflects an effort to reclaim an original approach to agricultural research and extension in which farmers were encouraged to work closely with researchers in identifying and solving problems in ways that were practical and feasible for farmers. Still, as Lockeretz and Anderson (1993) point out, we cannot assume that meeting farmers research needs will necessarily serve the purpose of alternative agriculture research, even at the farm level. They write that alternative agriculture programs have generally dealt with short-term problems, such as how to control a particular weed, which works against broader systems studies. The privileging of farmers in the movement for sustainable agriculture is similar to the current privileging of indigenous knowledge and peasant practices. Yet as Watts and Peet (1996) assert, "There is no pure, perfect, or easy solution waiting to be found" in the minds and practices of indigenous peoples. This also is true in the case of American farmers.

Gender and Power

Throughout history, gender has been a major determinant of who makes decisions, who controls resources, and who has their basic needs met. As Lipman-Blumen (1986: 54) frames the issue, "The paradigmatic power relationship between women and men, with its intransigent inequality mapped on all other relationships, across all nations, is the most crucial and fundamental issue underlying social justice." In every place in the world women are poorer, own less property, do more work, hold less power, are less educated, and suffer more hunger than men. Since gender is such a determining factor in access to and control of resources, it follows that gender relations are crucial to shaping the prospects for a more sustainable future (Rocheleau et al. 1996).

In American society, differences along lines of gender are arguably sharpest in the food and agriculture sector. In the U.S. agrifood system, including the majority of family farms, men control land, capital, and women's labor, and women's participation and interests have generally been subordinate to those of men (Sachs 1991, 1996). Of eleven major U.S. industries, agriculture has historically been the least likely to employ women as managers, executives, or administrators (U.S. Department of Labor 1989). Women employed in these positions make up less than one percent of the total managerial force in the agricultural industry. Women are also poorly represented among agricultural scientists. In fact, the agricultural sciences have proportionately fewer women scientists than do other scientific fields. Only 7 percent of agricultural scientists are women, as compared to 18 percent for all scientists (National Science Foundation 1989).

And even though the number of women-owned businesses in agriculture has almost doubled since 1980, only one business sector (the transportation, communication, and utilities sector) reports fewer women-owned businesses than agriculture. Of those who control U.S. farmland, only 4 percent are women (Economic Research Service 1985). Women tend to own smaller farms—the average size of farms owned by men is one-third larger than that of farms owned by women.

While women have always provided key labor on family farms and have become increasingly integrated into the entire commercial food system as wage laborers, that integration has segregated jobs along gender lines, with typically gendered pay scales. Worldwide, women's wages in agriculture are consistently lower than men's, sometimes as little as 63 percent of the male wage for comparable work (International Labour Office 1988). Despite two decades of affirmative action, American rural women earn significantly less than rural men. Irrespective of ethnicity, women are at an economic disadvantage to men in the rural work force. In 1987 less than a third of rural men workers, but more than half of women workers had incomes below the poverty line for a family of four, even though they worked the equivalent of a year-round, full-time job (Gorham 1992). Among U.S. agricultural laborers, women are more vulnerable to exploitation than men, and they are paid even lower wages and given fewer benefits than their male family members (Kearney and Nagengast 1989).

This double exploitation of women as both women and as workers (Fairclough 2001) is also evident in the domestic sphere. After all, it is in the home and the family where gender differences in the treatment of individuals and access to resources begin (Engberg 1996). Notwithstanding the increasing entry of women into the labor force, women remain overwhelmingly responsible for family food provisioning. Even where men share more of the domestic labor, food labor generally continues to be confined to women. Except where food is prepared and served outside the home, men are only marginally involved with food provisioning activities (Engberg 1996). Women continue to be the providers in terms of planning meals, choosing, shopping, and preparing food. There is an even darker aspect of the imbalance with socially accepted gender roles around food provisioning. Since men's needs "dominate the organization of cooking and eating" in terms of the composition and timing of meals, there can be serious repercussions if food is not prepared correctly and on time (Bell and Valentine 1997). In fact, the purchase, preparation, and serving of food has been found to be a key instigator of violent incidents in the home because it provokes some men's efforts to control the behavior and allocate the schedules of their domestic partners (Ellis 1983).

Hassanein (1999) points out that often the limitations based on gender faced by women in agricultural settings come not only from overt discrimination or institutional barriers but also from their socialization in rural communities and unequal gender relations experienced in daily life. In addition to material differences, agrarian ideology has shaped both men's and women's identities and experience of rural life. Traditionally, men have held the formal positions in organizations while women provide social cohesion (Miller and Net 1988, cited in Sachs 1996). This is also historically how women have participated in agrarian populist organizations, such as those directed toward saving the family farm and preserving farm programs. In the United States rural women have tended to join organizations that support their families or farm organizations rather than participating in organizations dedicated to women's empowerment (Sachs 1996). For example, one supporter of the California Women for Agriculture in the 1980s said that the women were involved basically on behalf of their men, from whom they get their ideas (Friedland 1991).

How are these kinds of asymmetries between genders dealt with in alternative agrifood movements? From the beginning women have played central roles in shaping and furthering alternative agrifood movements and institutions, including holding positions of leadership. For example, women lead the National Campaign for Sustainable Agriculture, the California Campaign for Sustainable Agriculture, the CFSC, the USDA Community Food Projects program, and the USDA SARE program. In their study of a sustainable agriculture group in Iowa, Peter and others (2000) found that women are better represented and more prominent in sustainable agriculture organizations than they are in conventional agricultural organizations. However, other studies have found less gender equality in sustainable agriculture movements. For example, a study of women in the sustainable agriculture movement in California found that while women were active in the movement, particularly at the grassroots level, men tended to hold the more visible leadership and decision-making positions (Sachs 1996). A Minnesota study found that while men were involved in such roles as teachers, leaders, and decision makers in the sustainable agriculture movement, women involved in the movement tended to occupy support roles such as providing food, working registration tables, and sending mailings (Meares 1997). While this imbalance is beginning to shift, historically men have been disproportionately represented in leadership roles such as project directors, conference speakers, and authors just about everywhere. Women have been correspondingly overrepresented in social cohesion roles such as organizing conferences, coordinating community endeavors, and fostering networks among different groups.

There are additional facets to this gender division of labor in alternative agrifood movements. One is the extent to which they may be increasing women's workloads. Some of the women mentioned in "Making the Transition from Conventional to Sustainable Agriculture" (Meares 1997) reported that their workloads had increased as a result of their partners' participation in the movement. Not surprisingly, men reported that they had benefited from participation in the sustainable agriculture movement, but women did not. In DeLind and Ferguson's (1999) study of community-supported agriculture, they discovered that women were the primary workers. It is possible that some practices advocated by alternative agrifood movements, such as farmers' markets and CSAs, can add not only to the work of farm women, but also to urban women's already overburdened workload in food procurement and preparation. Women's traditionally gendered home and farm responsibilities also can prevent them from participating in sustainable agriculture meetings and discussion groups (Meares 1997). This leads into the issue of gendered agrarian cultures.

It seems that while those working in sustainable agriculture adhere to less traditionally gendered relationships, in some cases traditional gender roles are reinscribed in sustainable agriculture. For example, Peter and others (2000) concluded that men who are interested in sustainable agriculture tend to be less controlling, more open to change and criticism, and better able to express emotions that men involved in conventional agriculture. On the other hand, they also learned that the category of "farmer" remained the exclusive domain of men's work, both in the eyes of the community and in the eyes of farm families. That is, both men and women saw the farmer as the man and the man as the farmer, while women were seen as farmer's helpers. How is it that these kinds of gendered roles prevail on sustainable agriculture farms given that those interested in sustainable agriculture are generally socially progressive?

In her study of sustainable agriculture networks, Hassanein (1999) captures the insidious, damaging ways in which agrarian women have been excluded from material and social power in America's agrifood system. One woman farmer reflects on the disabling experiences of never having had choices or controlling any resources: "Here you are thirty-five years old and still getting an allowance, as if you were a little kid. . . . You're an adult, but you never make adult decisions." In some arenas sustainable agriculture is based in concepts that preempt or run counter to a feminist agenda, that is to say, they contribute to the maintenance of gender-based status and power differences in the food and agriculture system. For example, the movement tends to glorify family farm and agrarian values without questioning the patriarchal privilege that underlies many of these values. In fact, the leader of the women's network studied by Hassanein was at pains to point out that it was not a "feminist" organization.[2] This kind of socialization helps to explain the relative silence of activists around gender issues in rural communities.

If the record on women's participation in sustainable agriculture is mixed, it is clear on the degree to which gender *issues* are addressed. Despite the highly gendered nature of the agrifood system, the amount of attention to gender relations in alternative agrifood efforts is negligible. Jackson (1994) uses the term "gender relations" rather than "feminism" and "patriarchy" to get away from the assumption of unitary interests among women and romanticizing women's "nature." With a few notable exceptions, research on alternative agrifood systems has ignored gender as an analytical category to the same extent as conventional agriculture research. Gender analysis emphasizes the importance of studying men and women in relation to each other at all levels of social organization. In the context of alternative food and agriculture gender analysis would involve studying how women have been excluded (or included too much), as well as examining the causes of and ways in which women are exploited, marginalized, and oppressed.

---

2. Two ideas are common to all strains of feminism: (1) that women are oppressed at all levels of society, and (2) that conscious political action is necessary to change this situation (Jackson 1994).

Although women are involved in sustainable agriculture practice and action, their involvement may be in fairly traditional roles in support of a movement and practice that remains relatively silent on gender issue or that reinforces existing gender inequities. Both passive reinforcement, by not directly addressing gender issues, and active reinforcement, through ideological constructs such as agrarianism, work against women's empowerment. Working toward gender equity would include equalizing power relationships—economic, discursive, and cultural—between women and men. Family farms, community, and localized food production may appeal to a nostalgia for an agrarian past, but they may also contain patterns that reinscribe long-standing oppressions of women. These patterns need to be identified, made visible, and acknowledged through alternative agrifood discourse and practice. This is a first step toward ensuring that existing gender inequalities in power, privilege, and opportunity are not unwittingly reproduced in alternative agrifood movements and the social forms they are creating and promoting.

Reflections on Democracy, Power, and Privilege

If the alternative agrifood movements are to achieve their environmental and social justice goals, they must democratize the allocation of opportunities and resources in order to give all people equal voice and agency. Progressive reforms can only be realized through the empowerment of those who are currently in subordinate positions (Hunter 1995). While involving those who have been excluded or subordinated is clearly required to meet liberal standards of democracy, it is in and of itself insufficient for achieving a deep democracy.[3] As Stiefel and Wolfe (1994: 5) point out, "After all, everyone 'participates' in society, whether as an effective actor or a passive victim."

Forms of liberal democracy such as inclusion of "stakeholders" and equal voting rights can only take us so far. Norgaard (1994) argues against the idea of governance by elected representatives and suggests that randomly and self-selected decision makers would instead provide greater opportu-

---

3. Hayward (1992) reminds us that although early Athens is held up as the essence of participatory democracy, women, men under twenty, slaves, and immigrants could not be citizens. For neither Rousseau nor Thomas Jefferson, were women part of the model or practice of democracy.

nity for everyone to participate. For example, in a study of two federally funded rural Enterprise Communities in West Virginia, the democratically elected board was all white, and women participated only in support roles, while the board picked by the executive director included African Americans and women in leadership roles (Maggard and Thompkins 1998). People tend to vote for people they perceive as having status and authority, based on cultural traditions. Without putting democratic content into democratic forms, a simple liberal democratic renewal will be insufficient to significantly change the food and agriculture system.

A call for democracy is empty without the concomitant democratization of economic and social power. The presumption that everyone can participate (much less equally) given current material and cultural circumstances is an illusion. Having "rights" does not necessarily mean being able to exercise them effectively (Sharp 1995). This is because social relations of power and privilege preexist discourse, institutions, and decision making. Social relations of power and privilege not only determine who is allowed to be part of the conversation but also shape who has the authority to speak and whose discursive contributions are considered worthwhile. The existing distribution of privilege and power compromise the possibility of authentic democratic participation. Power already accrues to some participants and not to others, and this power is determined by their institutional role and their socioeconomic status and gender or ethnic identity (Fairclough 1992). Full participation and autonomous agency to make decisions require the evening out of various forms of power in society.

Material Power

Material power is fundamental and basic. Control over resources lies at the very heart of all power relationships—between nations, socioeconomic and ethnic groups, generations, and women and men (Lipman-Blumen 1986). The maldistribution of material power limits authentic democracy. Just as equality in representation is essential for providing equal opportunities in decision making, so is equality in material circumstance essential for providing equal opportunities to have voice. It is simple—those with greater resources have greater power. Corbridge (1998) outlines Sen's premises on development,[4] which are germane to the issue of democracy. The premises are that:

4. See Dreze and Sen 1989 for a complete discussion on food security and the concept of entitlements.

- Development [and democracy] means the ability to make choices.
- The ability to make choices depends on assets.
- Having assets depends upon entitlements [both material, such as inheritance, and cultural].
- The distribution of entitlements is unfair.

Clearly, then, true democracy is impossible without equality of entitlements and assets. Lack of equality promotes a self-perpetuating cycle in which the marginality of people peripheral to decision-making processes is reproduced.

When one considers the existing distribution of resources in the United States, it is hard to imagine how people can participate equally in setting priorities and making decisions. For example, inequality in income distribution in this country is enormous. In the late 1990s the poorest 20 percent of families had an average income of $14,620, while richest 20 percent of families had an average income of $145,990—ten times as large (Bernstein et al. 2002). This acceleration in income inequality began in 1968 and has been increasing ever since. In 1994 the Gini index (a measure of income inequality) was already 17.5 percent above its 1968 level (Census Bureau 1996). The World Bank (2003) lists the Gini coefficient for the United States as 40.8, the same as Turkmenistan. (In a country where income distribution was equal, the Gini coefficient would be zero.) There is an even greater disparity in distribution of wealth than there is in distribution of income. And, of course, people do not have equal opportunities to acquire wealth. Having wealth is something largely outside a person's control. One's level of wealth is determined overwhelmingly by one factor—the level of wealth into which they were born (Keister 2000). "Birthright" and inheritance are not things one can acquire by working hard, getting a good education, or managing money wisely. If you are not born into wealth, it is incredibly unlikely that you will gain wealth.

Another form of material power is the way in which public and private funding shape priorities and projects in alternative agrifood movements and institutions. For example, food policy councils and community food planning efforts have tended to remain marginalized because they lack funds and institutional support (Gottlieb and Fisher 1996a). Dowler and Caraher (2003) report that food projects in the United Kingdom frequently have to reinvent themselves on an annual basis to avail themselves of funding opportunities. Similarly, they do not advocate fundamental changes in the agrifood system for fear of alienating their funders.

In the academic world universities increasingly base decisions about the value of a person's work on their ability to attract funding, which results in increased power and legitimacy for those who are able to do so. Exacerbating the problem is the effect of private donations on the priorities of the public agrifood research system. According to a former dean of the UC Davis College of Agriculture, private supporters can determine many agricultural research priorities by making relatively small financial contributions (McCalla 1978). Although public funds are used to cover the base costs (e.g., building, salaries, infrastructure) of the public agricultural research system, the actual research priorities have often been shaped by funding from private sources such as commodity organizations and agricultural products firms. Private, profit-driven funders are unlikely to fund research on the environmental or social justice issues that concern the advocates of alternative agrifood systems. Participation, action, and research cost money; yet essentially by definition money accumulates in social spaces that have flourished in the structure and culture of the existing social system.

This situation is confounded by the fact that most of those in leadership positions such as policymakers, academics, organization leaders, and even activists often come from and lead materially or socially privileged lives. This is not to diminish the importance of these actors to alternative agrifood efforts. They are absolutely essential precisely because they are in positions of power. What it means is that these privileges can make it difficult for them to feel the "raw nerve of outrage" (as Corbridge [1998] quotes E. P. Thompson) that comes from personally and consistently experiencing social injustice. Such conditions are primarily abstractions, not experiences, with corresponding implications for depth of understanding and commitment. What is more insidious is how holding a position of authority or prestige is regarded as an earned and purely individual achievement, while the social constraints on who can actually achieve these positions is ignored (Fairclough 2001). Anyone who has worked in academia can verify his observation about this reification of (primarily) class position. What is striking, says Fairclough (2001: 54), "is the extent to which, despite the claims of education to differentiate only on the grounds of merit, differentiation follows social class lines: the higher one goes in the educational system, the greater the predominance of people from capitalist, 'middle-class,' and professional backgrounds."

The conundrum lies in how to meet the needs of the voiceless and the most vulnerable within this negotiated space, both for the present and in the long run. Issues such as hunger and farm labor conditions are of concern primarily to those who have the least voice in the political economic system. For example, in many communities, the formation of food policy councils has failed where there was more emphasis on hunger than on other food-system issues (Dahlberg 1994a). The hungry are most often the poor, mostly women, children, ethnic minorities, and the elderly (Nestle and Guttmacher 1992). These are the people least able to participate in the current political system. Obviously, children cannot vote, but rates of voting are also low for the marginalized, oppressed, and underprivileged. The most disadvantaged and impoverished are often not able to participate in social movement committee meetings and actions. They may not have the time and resources to actively engage with social movements if they are already working long hours just to pay for food, clothing, and shelter.

Discursive Power

Beyond the asymmetries that correspond directly to income and wealth, there are other, more subtle distortions of democracy, such as differential access to discursive power. At all scales of decision making, the audibility of people's voices is modulated by cultural relations of power. Fairclough (2001) articulates the concept of cultural capital, pointing out that access to types of discourse and subject positions of power are "cultural goods" in the same way that wealth, income, and good housing are material goods. Both sorts of goods are unequally distributed, with the working class having substantially less of them than the professional, middle, and upper classes. Young (1995) points out that discursive social inequalities arise because ideal speech situations privilege some styles of speaking over others (e.g., reasoned argumentation versus stories of situated experience). Fairclough (1994) talks about democratization of discourse, by which he means the removal of inequalities and asymmetries in the discursive and linguistic rights, obligations, and prestige of groups of people.

People whose perspectives, ideas, and proposals get heard may be simply the most aggressive, loudest, and most confident, not necessarily those with the best ideas. This not only focuses attention on the viewpoints of these people, it simultaneously restricts the ability of others to present their perspectives. While this is not necessarily intentional, it is nonetheless damaging to democratic process. Indeed, as Gal (1992: 160) states, "The strongest form of power may well be the ability to define social reality; to impose visions of the world." Fairclough points out that one of the most powerful constraints on access is the way that having access to prestigious sorts of discourse and powerful subject positions enhances publicly acknowledged status and authority. There is a clear and potent connection between knowledge and power. Those who possess knowledge are better able to exercise power in ways that are more likely to bring about the changes they desire.

Gender constitutes a particular form of discursive imbalance. Goldring (1996) points out that even when rural people's voices are listened to, it is the men's voices that receive privileged attention. A study of the Women, Food, and Agriculture Network reported that while the women saw progressive farm organizations as allies, their experience in these organizations had often been that women's voices were silenced rather than amplified (Wells 1998). This imbalance obtains even in "enlightened" intellectual settings. For example, studies have shown that men start to feel that women are dominating when their participation rate reaches 20 to 30 percent in settings such as classrooms, meetings, and seminars (Thornborrow 2002). Another study of an academic discussion list on the internet found that there is a "threshold" of women's participation that men will tolerate (Herring et al. 1995). This is particularly troubling given the reluctance with which many women even dare to speak up, given the cultural traditions that effectively silence them. Even when women's voices are heard, they are often marginalized and stripped of any real power. For example, the Second International Conference on Women in Agriculture was held to "elevate women's voices in a predominantly male world." Yet a request by a group of women to develop a session to work on resolutions and strategies for change was rejected by the conference convener (Women Food & Agriculture Network 1998).

These asymmetrical distributions of power, status, and privilege—seen or unseen—make it clear why a simple form of democracy in which a diversity of voices are included is insufficient to meet democratic ideals of equality in priority setting and decision making. For example, a study on participation in development found that while the poorer people were being included in discussions, they were still excluded from gaining any control over decision-making and regulative institutions (Stiefel and Wolf 1994). At a workshop on increasing diversity within the cfsc conference, participants pointed out that they did not simply want to be *included* in an existing framework; they wanted to be full participants in *creating* the framework that established priorities and strategies. Inclusion is also insufficient to recognize or reconfigure systemic relations of power in the agrifood system.

For example, while the Kellogg IFS network includes projects designed to benefit African-American growers, it stops short of challenging the structures that have consistently discriminated against them. These are honest efforts to bring excluded people into the fold, but they do not go so far as to challenge the institutions and cultures that have systematically excluded people. Participation and empowerment are dialectically related. People who participate are empowered; empowered people participate.

Alternative agrifood movements and institutions are committed to the principles and the careful practice of democracy and are much more democratic that the agrifood system has been in the past. However, there is some distance to go in equally valuing the needs of all participants in the agrifood system. The movements and institutions are also working in a historical and cultural milieu of differential distributions of power and privilege. While conscious efforts are being made to overcome this, it will not be easy or happen quickly. Similar problems of power and privilege are embedded in place and community. While the ideals of democracy and empowerment are crucial as general principles, alternative agrifood movements need to address a perennial question, posed in a more general context by Miller and others, "*Whose* empowerment to do *what*?" (Miller et al. 1995: 121).

7	politics of complacency?
	rethinking food-system localization

Like many other contemporary social movements, alternative agrifood movements have focused increasingly on discourses and strategies of localization. These days almost all alternative agrifood movements promote food-system localization as both a concept and a strategy. The reason for this emphasis on local food systems is both practical and political. On the practical side is an interest in reducing energy costs used in transporting and storing food. The political aspect is based on interest in deepening democratic principles and practice. This turn toward the local by social movements began in the 1960s, based partly on ideological commitments to participatory democracy, decentralization, and human-scale systems, and partly on practical limitations in resources (Flacks 1995). The idea of local food systems, and localization more generally, is appealing on many levels. At the same time, it warrants closer examination of the assumptions that undergird food-system localization efforts.

While localization has proved to be a popular, galvanizing discourse and an effective organizing strategy, it may be worth taking a step back and more carefully considering the premises and implications of localism. In this chapter I raise questions about the role of localist discourses and practices in achieving the democratic and social justice goals of alternative agrifood movements. These are the kinds of issues I want to engage in discussing localism and alternative agrifood movements. I am setting aside practical questions such as the possibility of providing an adequate diet based on local foods or the role of comparative advantage in economic development. My primary question is this: To what extent can the turn toward localism *ipso facto* increase participation and social equity in the agrifood system? This chapter begins by presenting the case for local food systems and reviewing the pace of globalization that has resulted in an increased interest in localism and community during the 1990s. It then turns to questions about assumptions about democracy and power within communities and the meanings of community and place. My hope is that the politics of place does not devolve into the politics of complacency, where some people flourish in a vacuum of concern for others.

The Case for Local Food Systems

In lieu of a globalized, depersonalized agrifood political economy, alternative agrifood movements advocate the development of decentralized and community-based local food systems. Community food security, for example, is "squarely within the anti-globalization community," and groups are working to develop "concrete alternatives that promote locally grown foods instead of globally sourced ones" (Fisher 2002b). The ideas that "place matters" and "scale matters" have been central to the community food security approach (Gottlieb and Joseph 1997: 13). According to Feenstra and Campbell, community food projects are collaborative efforts to "integrate agricultural production with food distribution in order to enhance the economic, environmental, and social well-being of a *particular* place" (1998: i, italics added). Alternative agrifood movements also support shortening the distance between farmers and consumers and the miles that food travels from its point of production to its point of consumption. According to Fox (1990: 729), one of the most frequent themes in the sustainable agriculture movement is that "trade *per se* is inherently environmentally degrading."[1] Arguments in favor of food-system localization include strengthening community markets for local farmers and food processors, reversing the decline in the numbers of family farms, creating local jobs, reducing environmental degradation, and protecting farmland from urbanization pressures through rural economic development, fostering direct democratic participation in the local food economy, and cultivating caring relationships among people in a community (see, for example, Dahlberg 1994b, Feenstra

---

1. This is not always the case, however, as many U.S. organic farmers are looking to capture price premiums through the European and Japanese markets.

1997, Henderson 1998, and Kloppenburg et al. 1996). In recent years, numerous local food-systems projects have been established throughout the United States. More than half of the alternative agrifood organizations we studied in California were involved in the development of local food systems. This localization and decentralization are seen as sources of vitality and strength for the alternative agrifood movement (Henderson 1998).

Local food systems are seen not only as necessary to achieve sustainability and food security, but also as *embodying* these qualities. Feenstra (1997: 28) articulates a viewpoint widely shared in alternative agrifood movements in stating that local food systems "use ecologically sound production and distribution practices, and enhance social equity and democracy for all members of the community." Local food systems are considered to manifest high levels of noninstrumental, caring, and ethical social relations (Hinrichs 2003).

It has become "common sense" in alternative agrifood movements that achieving sustainability and food security requires transferring power and responsibility from central government to local government. As a participant in the CASA project writes, "A strategy that links sustainable agriculture to community-controlled economic development is consistent with the movement's widely shared goal of promoting food and agricultural systems that are environmentally sound, economically viable, and socially just" (Campbell 1997a: 37). The vision of community food security is based on "the recognition that communities need to be empowered to become food secure" (Gottlieb and Joseph 1997: 5). Poppendieck (1997: 175), a scholar of antihunger movements and programs, states that in order to solve problems of hunger and undernutrition, "community development and locally based solutions are fundamental." The assumption underlying these approaches is that at a community level people will make more appropriate and compassionate decisions than those outside the community and that processes at a community level will be equitable. This is argued by Kemmis in his influential book, *Community and the Politics of Place*.

Globalization and the Politics of Place

The attraction of local frameworks and community empowerment has increased recently because of changes in political-economic circumstances, including the rapid expansion of the global market in the 1980s. Illustrating this trend is the fact that by 1998, fifty-one of the world's largest one hundred economic entities were corporations (Barlow and Clarke 1998), organizations in which very few people play a role in setting priorities and making decisions. The World Trade Organization (WTO), which sets the rules of international trade, codifies the move away from democratic decision making and toward increasing influence of private businesses. The WTO is an international body that makes decisions that affect everyone in the world, yet there is little opportunity for public participation in its discussions. In its decisions in trade disputes, economic growth has consistently trumped environmental, health, and labor objectives. This is a significant change from how economic and social goals have been balanced within Western national governments since the 1930s.

The proposed Multilateral Agreement on Investment (MAI) went even further than the WTO. The MAI would have given corporations a legal status similar to that of nation-states while reducing the power of nation-states over corporations to enforce labor and environmental standards. The MAI was much more far-reaching than NAFTA in that it involved more countries (twenty-nine OECD nations), was legally binding for a longer period of time, and placed more extensive bans on performance standards. Sessions discussing the MAI were the antithesis of democracy. Not even U.S. congressional committees with jurisdiction over international commerce were briefed about MAI negotiations even though negotiations involving U.S. transnational executives had been ongoing for three years (Barlow and Clarke 1998). Labor, environmental, and citizens groups were not consulted; they did not even know the MAI existed, since negotiations were held in secret.

While governments do not by any means have excellent records of meeting the social needs of their populations, many have at least offered some measure of protection from the worst effects of the market system through Keynesian economics and social and environmental regulations. In recent times, however, the role of governments in ameliorating the negative consequences of the free market system has been reduced significantly. For example, nearly every advanced industrialized country has shed much of its responsibilities for social welfare. In five of these countries—the United States, the United Kingdom, Canada, Australia, and New Zealand—governments have been changing welfare systems while neglecting the growing issue of hunger and food insecurity (Riches 1997). In the United States, the 1996 changes in food programs for the poor represent the largest cutbacks since these programs were first established. National governments are now functioning less and less as buffers between the economy and society, and more and more as agencies for promoting the growth of global capital. This weakening of the nation-state vis-à-vis capital and the abrogation of state responsibilities for social welfare have meant that people and the environment are increasingly vulnerable. Traditionally unequal power relations and distributions of resources have risen to new levels of disparity through this transnational political-economic restructuring.

The retraction of the state creates a vacuum of effective political remedies for worsening environmental, economic, and social problems, and the turn toward localism has emerged to fill the void. Localism provides a defensive position against the disempowering and homogenizing effects of globalization. People turn to local issues and local activism as a way in which they can experience empowerment, as an antidote to despair (Young 1990). In today's complex, fast-paced world many people crave face-to-face contact and situations in which they feel they can make a difference.

To an extent, all social movements are local or at least have local manifestations. General efforts at social change are always made up of particular, local efforts. Things are "done" in concrete spaces that can only be local. There is strength in decentralized, particular, local movements, both in and of themselves and as components of broader social movements. Local movements or local manifestations of broad-scale movements have been very effective in the past. For example, the environmental movement gained its strength, not from the work of national organizations, but from local protests and struggles over issues such as land use and toxic waste disposal (Flacks 1995). In addition to successfully winning some of their struggles, these local efforts enabled those who had not considered themselves activists to acquire the knowledge and skills to continue to organize, some of whom went on to seek electoral office.

Local efforts provide the opportunity for direct participation, in many cases for people who would not participate in national or statewide efforts. Local organizing can also be community building and empowering because people can effect changes that can be measured in visible, tangible benefits. Local efforts can be embraced and acted upon sooner and more fluidly than those at larger scales. These local developments are necessary proving grounds for creating and troubleshooting alternatives that can shape the future. They are examples of the preferred strategy of Kloppenburg and others (1996). They advocate "secession" from the dominant agrifood system (by creating alternatives to it), followed by the "succession" of these forms (that is to say, gradually transferring resources and commitments from dominant to alternative agrifood institutions). They also point out that working toward what may seem like only reforms can create resources and ideas for deeper and more far-reaching changes.

Globalization is as rampant in the agrifood sector as in any, beginning perhaps even earlier than in most. While agriculture has always involved some form of trade, in the past decades, changes in the agrifood sector have blurred national boundaries, intensified agricultural specialization for both enterprises and regions, and created large agroindustrial complexes (Friedmann and McMichael 1989). This has led to phenomenal and accelerated concentration of economic power. For example, while in 1997 the top five agrifood firms accounted for 24 percent of retail sales, by 2001 that figure had almost doubled to 42 percent (Hendrickson et al. 2001). Through this process, food has become both distant and durable as particularities of time and space are compressed in its production and consumption (Friedmann 1994). As a result people have increasingly lost control over the source and quality of their food, leading to alienation from the agrifood system.

Decision Making Within Communities

The emphasis on local politics is related conceptually to the approach of social ecology (see, for example, Biehl and Bookchin 1998). From this perspective, the state can never act other than as impersonally bureaucratic and alienating, failing to meet the needs of real people. The alternative is a local politics grounded in the municipality, the "natural" site of social, political, and environmental change (Roussopoulos 1993). For Atkinson (1992: 216), environmental sustainability depends upon local control "because that is manageable in terms of the physical wellsprings of human consciousness, namely a manageable number of humans participating in the social and natural metabolic process." Campbell (1997a: 43) agrees: "When the focus is community problem-solving rather than the inherently adversarial budgetary and regulatory processes of government, greater opportunities exist to integrate environmental and social concerns."

The notion that local communities will make better decisions than larger, more distant polities about food systems is based on an expectation of fluid cooperation among groups with quite different material interests. For example, Lezberg and Kloppenburg (1996) write that a key problem with the agrifood system is that people have lost control over decisions made within their locale, which are instead made by those who have little interest in those who live in the community. The premise that those within communities will make more equitable decisions rests upon the assumption that everyone acting within these groups has more or less equal power. For example, as Campbell (1997a: 43) writes, "A shared commitment to place and the expectation of continuing encounters tend to check behavior that deviates from shared community interests and to subsume separate issues under a broader concern for the community's welfare." But what are "shared community interests" and how are they decided? A community can never be completely homogenous in its goals, given that different social actors possess different material and cultural resources as a result of the social spaces they occupy as producers or consumers, men or women, rich or poor, and so on. For example, in political struggles over New Deal programs wealthy farmers opposed tenant farmers and farmworkers over such core class issues as ownership, profit, wages, and unionization (Domhoff 1991). If interests were community-based, community members would have allied with each other along geographical lines, rather than as they did, along class lines. Localism subordinates differences to a mythical "community interest."

It is not the case that reducing the scale of decision making will necessarily enable excluded people to have voice and power that they have not been able to exercise at higher levels of decision making. Local communities embody the same kinds of power asymmetries present at the national level and may even magnify them. For example, at the local level gender imbalances in agricultural decision making become even sharper than at the federal level. Of the local officials who run powerful USDA county and state committees 98 percent are European-American men (*Kansas City Star* 1991). As Lewis (1992) points out, local politics are just as likely to be dominated by "grasping oligarchies" as by "equality-minded citizens' councils."

More participatory democracy at local levels is absolutely necessary to work toward an environmentally sound and socially just agrifood system, but it in and of itself is not sufficient because some voices drown out others. A study designed to examine changes in food-system perspectives as a result of a participatory planning process found that engagement in the highly participatory process was associated with "*decreased* salience of social justice and environmental concerns and *increased* salience of a viewpoint that is unsympathetic to these concerns" (Pelletier et al. 2000). The researchers observed that, contrary to common perceptions, participatory or collaborative approaches involving diverse stakeholders could possibly narrow—rather than expand—the range of values considered. I have had similar experiences participating in collaborative projects and planning processes. What tends to happen is that people want to pursue the paths of least resistance, often choosing and pursuing priorities that are "normal" and noncontroversial. Sharp (1995) observes that powerful people within a community can easily subject others to cultural, political, and economic domination.

Power Within Places

There are clear asymmetries of power and privilege embedded within small communities. Geographical proximity does not reduce social and economic distances among people. Communities may—and generally do—harbor large social disparities. For example, there are greater inequalities in income and food consumption within regions than among countries (World Bank 1986). Disparities in wealth, power, and privilege fall more along lines of class, ethnicity, and gender than along spatial boundaries. Ethnic segregation is deeply inscribed in rural areas. Almost half of rural America's minority population lives in counties with high concentrations of minorities —counties that also tend to be geographically isolated and suffer from generally high rates of poverty (Cromartie 1999). However, poverty rates are differentiated by ethnicity. In predominantly black and Native American counties, poverty rates for these groups reach nearly 50 percent while the poverty rate for whites is only 13 percent. In California's central coast, strawberry workers are concentrated into poor neighborhoods; the poverty rate in one Watsonville neighborhood in which many strawberry workers live is twice the national average (United Farm Workers Union 1996).

Segregation in agrarian communities is also by occupation, with resultant differences in resource and power allocations. Although African Americans, Latinos, and Asian Americans have provided much of the farm labor in U.S. agriculture, they are much less likely than European Americans to be farm operators. Nonwhites comprise nearly 25 percent of the population, yet ethnic minorities operate a mere 2 percent of the farms in the United States (Census Bureau 1987). Even in a state as ethnically diverse as California, where 43 percent of the population is nonwhite, less than 7 percent of farm operators are nonwhite (Census Bureau 1987). In contrast, California's farm labor force is composed almost exclusively of ethnic minorities, 95 percent of whom are foreign born (Kuminoff et al. 2000).

Reducing the scale of human interactions does not necessarily achieve the social equity or empowerment espoused by alternative agrifood movements. Small-scale institutions are not always more equitable or desirable. A survey of California farmworkers, for instance, found that the majority preferred to work on large farms rather than small farms because they experienced fewer abuses and received higher wages on the large farms than on the small ones (Buck et al. 1997). Another small-scale institution is the family, yet one in which terror and violence may reign. For an American woman, the most dangerous place she can be is in her home, the most "local" of places. Rates of reporting of domestic violence in rural communities are low because it is more difficult to maintain anonymity and because the women are economically tied to the farm.

In many instances small communities are anything but liberatory for those traditionally marginalized. In rural minority counties, there are even further separations by race and ethnicity at municipal and neighborhood levels, which generally involves relative economic disadvantage for these groups (Cromartie 1999). Localism tends to be positively associated with conservative positions on social issues and negatively associated with concern for social justice. Tightly knit communities can be constricting for those who do not fit community norms (Bell and Valentine 1997). These communities can place sanctions on nonconforming behavior such as mixed "race" or same-sex partnerships, with communities serving as loci of social control for what is perceived as deviant behavior. While there are abundant assertions that face-to-face relationships are purer, more authentic, and more just than relationships mediated across space and time, evidence suggests this is often not the case.

Working only at the local level is not only insufficient to rectify inequities, localism may actually be the source of these inequities. In many cases the disenfranchised have turned to the federal government for relief precisely because progressive change was impossible at the local level or because local elites persisted in denying them basic rights. For example, in the South it took national legislation to overcome local preferences for racial segregation. "The realities of southern power dictated that organizers had to do more than promote salience and efficacy at the grassroots. Change necessitated intervention from the North" (Goldberg 1991). When studies in the 1960s found that rural communities held deep pockets of hunger, action to alleviate hunger was taken at the federal—not local—level. In these cases, local control resulted in the antithesis of social justice.

Existing power relations and dependencies can preclude the spontaneous generation of participation, which may require outside stimulus (Bryant 1991). Consider the well-known Indian Chipko movement, often held up as the quintessential local, grassroots environmental movement. Chipko became a regional autonomy movement, but was shaped by earlier Communist Party and Gandhian organizing and, therefore, cannot be considered a purely local or indigenous movement (Watts and McCarthy 1997).

Despite the priority given to it in progressive social movements, local, grassroots organizing is not always progressive in intent. For example, high-income groups organize to protect their particular ways of life, neighborhood residents organize to keep out low-income housing, and right-wing groups organize antitax initiatives and lobby for censorship of library materials. While the premise and hope in alternative agrifood movements is that local communities are democratic and progressive, they can clearly also be exclusionary and regressive. At the same time, Miller, Rein, and Levitt (1995) point out that participation in local actions rarely has a deep impact on people's political attitudes or brings about significant social change. The other side of the power of local efforts is that local successes can also lull activists into an isolationist complacency without their having accomplished anything of lasting significance. This ambivalence greatly limits the potential of grassroots organizing to democratize American life.

Related to a preference for local decision making is a perspective that locals know more than outsiders do. Pretty (1995), for example, writes that sustainability is only possible with local inputs, local control, and local knowledge. Certainly people at the local level have insights and knowledge not available to "outsiders," and this knowledge is embodied in their practices. However, just like "scientific" knowledge, local and indigenous knowledge can only ever be partial and particular. All knowledge is in context —situated in time, place, and most important, social relations. Knowledge tied to a "place" is not homogenous but differentiated by divisions of labor and power, which are in turn differentiated by class, gender, and race (Feldman and Welsh 1995).

There are things that depend upon local knowledge, such as knowing the best time to plant a certain crop. At the same time, there is necessary information that local knowledge cannot provide, and large-scale, global efforts can bring home issues that had previously been imperceptible. As Harvey (1996: 303) points out, place-based knowledge is "insufficient to understand broader socio-ecological processes occurring at scales that cannot be directly experienced and which are therefore outside of phenomenological reach." An example of an international focus that highlighted issues at home is found in the United Students Against Sweatshops (USAS), a growing student movement organized originally to improve labor conditions in the overseas garment industry. Although the movement tends to be composed predominantly of privileged white students (even more than other student movements of the past), the activists focus on everyday inequalities, gaps in wealth, and the lack of democratic accountability (Featherstone and United Students Against Sweatshops 2002). While many activists became engaged because of injustices in areas far from home, increasingly the USAS movement is promoting living-wage campaigns, working with domestic labor unions, and working on racial justice issues in their own communities.

Despite the appeal of primarily local, endogenous organizing in social movements, this may not always be possible or desirable. Too much focus on the local can lead to a lack of wider-scale organizing. For example, while regional and statewide organizing in community food security is taking place in many places around the country, these activities are "neither linked to each other nor are they products of a more systematic approach by the national organization" (Gottlieb and Joseph 1997: 2). An analysis of local food projects in the United Kingdom found that local food projects could not address changes needed in economic structures and that the issues of living on a low income were often overlooked in the search for quick solutions (Dowler and Caraher 2003).

More participatory democracy at local levels is absolutely necessary to the success of the sustainability and community food security movements, but local politics has to work in conjunction with, not instead of, national and international politics. In working toward food security and sustainability, some analyses and actions will need to remain local; others will need to be national or international in scope. It will be equally important to clarify what types of food security and sustainability each level can realistically address and achieve. Local struggles always connect to systems and issues outside the locality; to be effective they must also connect to larger-scale efforts.

Power Among Places

A latent analogue of the growth of the emphasis on the local is a type of localism that can become defensive and protectionist. Defensive localism involves reducing federal spending, pushing responsibilities down to lower levels of government, and attempting to contain social problems within defined spatial and political boundaries. The politics of defensive localism has been a key feature of the politics of race and poverty since the early 1980s (Weir 1994). Antihunger activists struggled against this trend by working to stop block granting of federal food programs in the 1996 welfare legislation. They reasoned that local control would lead to lower levels of food assistance, and the assistance that was available might be provided more sporadically or preferentially in the absence of federal standards. For example, welfare reform in the late 1990s showed how decentralization can mean increased costs without increased power to change the conditions that gave rise to the problems in the first place. The politics of defensive localism makes localities responsible only for the problems that occur within their jurisdictions in a world that is already defined and fragmented by income and ethnicity. "Local empowerment can become a very conservative goal that allows the broader political community to concentrate social and economic problems in particular places and refuse to take responsibility for those problems" (Weir 1994: 341).

Local actions may produce unintended negative effects such as the export of garbage and toxic waste to other areas. One form of defensive localism is the "not-in-my-backyard" (NIMBY) movements of the 1980s that worked to keep toxic waste out of their communities. The logical implication would be that the waste would then be dumped in communities with less power to resist. Localism may bring about marginal defensive actions that can pit groups against each other. For example, the leaders of a community food security project were proud of its success in reducing food imports from outside the locality. They were uninterested, however, in the impact this localization may have had on the livelihoods of those who depended on the previous arrangements (e.g., produce truck drivers or nonlocal small farmers). This strategy is another kind of defensive localism, refuting the notion that low incomes, not outsourcing, are the cause of food insecurity. This kind of localism enables communities to "do what powerful economic elites already do—displace costs onto others" (Hunter 1995: 337). Business has long used spatial dispersal and competition between places as a major strategy for reducing labor costs and ensuring a docile, compliant workforce.

The emphasis on community and localism also raises the question: Who is not "us"? Localism can be based on a category of "otherness" that reduces the scope of whom we care about. Community can be defined as *against* others and thus be exclusionary. Being part of a community necessitates defining others as outside that community. This is how social cliques function among children. Is the notion that boundaries such as farms, communities, or nations can or should serve as the boundaries of concern ethically defensible? Is it possible to just protect our community from food insecurity without protecting everyone? Would we want to? Can alternative agri-food strategies and solutions be defined in terms of "*only*-in-my-backyard?"

For Kloppenburg and others (1996) the idea of a global foodshed is an oxymoron. I am not so sure. It is both possible and ethically required that we work to develop an agrifood system that enables everyone to share the bounty. For the ethicist Peter Singer, geographical distance should in no way absolve people of responsibility to ameliorate suffering elsewhere. The CFSC has taken this tack, and in 2002 formed a committee to look at food security issues on a global level. In Milio's (1991) conception of an equitable nutrition policy, costs and benefits of nutrition are shared equitably among people and localities, with an equitable geographical distribution of healthy food.

As a result of historical processes, communities vary widely in the resources they can bring to bear in developing sufficient, sustainable, and equitable agrifood systems. Not all communities have equal power for a number of historical political-economic reasons. For example, California has a powerful economy because of a fortuitous abundance of natural resources and its importation and exploitation of workers from other countries. African nations, on the other hand, tend to have weak economies, in large part because the colonial powers looted their resources, enslaved a significant portion of their labor force, and converted their most fertile land from basic food production to large-scale commercial farming. Prior to colonization, Africa was self-sufficient in food, often producing large surpluses, at a time when many Europeans went hungry (Rau 1991). The localist perspective tends to elide the effects that the ravages of history have on contemporary conditions. For example, U.S. counties with high concentrations of low-income African Americans are coincidental with former plantation areas in the South (Cromartie 1999). The differential distributions of power among places compromises the defensibility of a purely localist perspective. While alternative agrifood movements do not advocate protectionist or defensive localism, this remains a logical corollary and possible consequence of the focus on locality.

Meanings of Community and Place

In alternative agrifood discourse, community, place, and local tend to be reified—treated as if they are things in and of themselves rather than social ideas and inventions. Advocates for local food systems speak in terms of bioregions, watersheds, and site specificity. For example, Herrin and

Gussow (1989) eschew the idea of using political boundaries (such as counties) to develop local food systems because they are not defined by bioregional delimitations such as watershed, soil types, climate, flora, and fauna. This raises the question of what constitutes place. While some aspects of life are more or less determined by locale and climate, social relationships are not. As much or more than sets of physical spaces, places are socially constructed circuits of geographically bounded social relationships that have been shaped through interactions with other places. Harvey (1996: 320) points out that to talk about the "power of place" as if places "possess causal powers is to engage in the grossest of fetishisms" unless we clearly define place as a social process.

Those in alternative agrifood movements often speak of "reembedding" production and consumption in local communities (see, for example, Dahlberg 1994a, Duncan 1996, Friedmann 1993, and Kloppenburg et al. 1996). While globalization has clearly accelerated in past decades, this idea of "re"-embeddedness may be somewhat ahistorical in the American case, reaching back to a time that perhaps never was. The history of U.S. agriculture after European contact is one of a distinct and purposeful "disembeddedness" of production and consumption, making it something of a myth that local, agrarian economy flourished in the United States. The development of U.S. agriculture depended on mass immigrations onto land made available by the eviction, and decimation, of its indigenous inhabitants. From the beginning these farmers produced goods for the world market, growing staple foods more cheaply than was possible in Europe. In fact, the objective of the first American agricultural policy was to speed the expansion of commercial farming and the export of agricultural products (Danbom 1997).[2] O'Connor (1998: 302) points out that "localities define themselves (or acquire self-definitions), both cultural and environmental," in ways that are constituted by global relationships. Agrarian communities are agrarian precisely because of the global demand for grains, for example. Ranching communities raise cattle because of outside markets, often on land owned not locally but by the federal government. Local, regional, national, and international are related dialectically; they all condition and shape each other.

---

2. The production of agricultural surpluses was seen as a necessary part of the nation's development process and agricultural exports were crucial for the economy as a whole, not just for farmers.

In alternative agrifood discourse community can often be reified rather than treated as a contingent social construction. The argument for localism rests on the premise that neighborhoods, cities, and towns are more authentic arenas of public life than those operating on a larger scale. Yet community has no practical meaning independent of the real people who construct it and act in it. It is defined differently by different people as mediated by income, wealth, property ownership, occupation, gender, ethnicity, age, and many other personal characteristics. Neither is community necessarily tied to place or identity. For example, migrant agricultural workers belong to what Kearney and Nagengast (1989) term a "transnational community," that is, one that does not fit conventional local, sociospatial conceptions of community. In this and in many other situations, identity is not necessarily tied to the place in which one resides. Location is not the only, or even primary aspect, of identity, which is constructed across many different dimensions, the most obvious of which are social class, gender, ethnicity, and sexual orientation. And these categories have variations within them as well. Communities and places are shaped both by global dynamics, and through social relationships of power and privilege embedded within the place itself. A reified or romanticized notion of community can obscure very real and problematic class, gender, and ethnic divisions within societies, communities, and households. At the level of the local community many forms of exploitation and oppression can still flourish.

Globalization produces winners and losers, but so does localism. Differences are socially produced in spaces and in places. Globalization is not just something that "happens." For centuries it has been "a specific project pursued and endorsed by particular powers in particular places that have sought and gained incredible benefits and augmentations of their wealth and power from freedoms of trade" (Harvey 2000: 81). Local social systems are often comprised of similar economic and political dynamics. At both global and local scales those who benefit—and those who do not— are arranged along already familiar lines of ethnicity, class, and gender. Koc (1994) suggests that "globalization" become a term for the knowledge that we share the same world, which requires responsible and caring relationships among members of the world community. This makes sense both ethically and strategically, given differential distributions of resources and power among localities. Working at the local level is important, but insufficient to developing environmentally sound and socially just agrifood systems. This is because, as Harvey (1996: 353) observes, "The contemporary emphasis on the local, while it enhances certain kinds of sensitivities, totally erases others and thereby truncates rather than emancipates the field of political engagement and action." Assertions about the value of local food systems can be confirmed through empirical research. In the meantime, there are reasons to be circumspect about advocacy of the local without directly addressing issues of participation and equity.

8

the politics of sustainability and sustenance

While alternative agrifood movements tend to focus on the individual and the community, they are also active at the formal, institutional level. They need to work at this macro level because of the determining role of public policy in configuring the agrifood system we have today. As Milio (1981) puts it, "At every step in the contemporary food chain, federal policy shapes the likelihood of what food producers will or will not do, and what people will or will not eat." The government has clearly played a major part in creating the social and environmental conditions in the contemporary agrifood system—both the positive and the negative, briefly described in first section of this chapter. This same state, with democratic participation on the part of alternative agrifood movements could develop effective policies and programs for sustainability and sustenance.

Alternative agrifood movements have worked tirelessly to broaden the goals of and increase participation in policymaking around food and agriculture, directly engaging the federal political process for the past two decades. Alternative agrifood movements have introduced and expanded concepts of sustainability and food security within powerful agrifood institutions in the United States. The movements have made great strides in changing—or at least diversifying—the perspectives and methods within traditional agrifood institutions such as the USDA. It is remarkable how much has already been accomplished, given highly constraining political and economic environments. In this chapter I review some of the ways in which federal agricultural policies have been inconsistent with the environmental and social priorities of alternative agrifood movements. I then discuss some of the reasons for this, such as inertia and the lack of democracy in the policymaking process. Creating policies that are consistent with environmental and social priorities presupposes alliances between diverse social movements and within alternative agrifood movements. While much progress has already been made in this direction, there are challenges in forming alliances that need to be taken into account. Still, successful connections are being made between social and environmental movements on local, national, and international levels; some of these are highlighted at the end of this chapter.

Configurations of U.S. Federal Food and Agricultural Policy

Food and agricultural policies and the political economy from which they flow have constituted the most dominant force in shaping agriculture over the past half-century (Clunies-Ross and Hildyard 1992). The official goals of U.S. food and agricultural policy seem reasonable enough. They have included providing an adequate and secure supply of food at reasonable prices, stabilizing farm incomes, ensuring consumer safety, preserving family farms, and more recently, conserving resources. But how well have these policy objectives been met? This section addresses the effects federal food and agricultural policies have had on economic democracy.

Environment

It is well known that federal farm programs have contributed to environmental problems in agriculture. For example, fertilizer and pesticide pollution, soil erosion, and groundwater depletion have been driven at least in part by commodity support programs, especially those concerned with food and feed grains, cotton, and sugar (Young 1989). These programs affect production decisions on two-thirds of the harvested cropland in the this country (NRC 1989). Appearance-driven grading procedures for produce also encourage the use of pesticides. Thus, agricultural resource conservation programs are opposed in practice if not intent by other governmental programs. While the omnibus farm bills since 1985 have included more environmental provisions than any previous farm bills, these have been mostly marginal additions to the overall entrenched set of traditional farm policies. Overall, American food and agriculture policy has been skewed toward business priorities, not environmental soundness or consumer health.

Food and Health

In policies related to food and health, the government has seemed to privilege the preferences of industry over those of social well-being. A case in point is the development of the food pyramid, intended as a national nutrition education graphic. The USDA spent almost a million dollars to assure meat and dairy interests that the food pyramid graphic did not present an economic threat to their industries, despite the fact that many of the major American diseases are related to consumption of these food groups (Nestle 1993). Still, the original graphic, which depicted a lower number of daily servings of meat and dairy than what is shown on today's food pyramid, was withdrawn because of persistent industry protests. Apparently, policymakers were unconcerned with the negative reaction of the nutrition community to this withdrawal because the nutrition community is comparatively small relative to the meat and dairy industry and lacks their political influence (Sims 1998). Similar conclusions about prioritizing business over health in food and agriculture policy can be drawn in the way that pesticides are regulated. With the advent of new synthetic chemical pesticides, the Federal Insecticide, Fungicide, and Rodenticide Act of 1947 was passed, requiring that pesticides be registered before going on the market and that packages be labeled with their contents. However, administration of the law fell to the USDA, which concentrated on pesticide efficacy rather than health effects. As a result, few highly toxic chemicals were banned (Blanpied 1984).

Cheap Food?

The objective of providing an adequate supply of food at reasonable prices has always been a top priority in American food and agricultural policy. Indeed, a primary justification for production-oriented farm programs has been to keep food costs low for consumers—a "cheap food policy" (Browne et al. 1992). Certainly, American agriculture has achieved production levels sufficient to provide adequate food for everyone. Whether this food is "cheap" or not is a different question. For example, between 1982 and 1988, higher food prices resulting from restricted supplies generated by farm programs cost U.S. consumers between $5 and $10 billion in indirect costs (Faeth et al. 1991). Studies by agricultural economists and the President's Council of Economic Advisers have demonstrated that farm program costs to consumers and taxpayers exceed producer benefits by several billion dollars per year (Luttrell 1989). Congress has regularly voted to maintain supports for sugar and tobacco programs despite how skewed their benefits have been, both among producers and between producers and consumers. For example, in the 1991 U.S. sugar program, 58 percent of the benefits went to one percent of sugarcane growers; this program costs American consumers $1.4 billion a year and taxpayers $90 million a year (Hamel 1995). Given that low-income people spend a larger portion of their incomes on food than do middle- or high-income people and that they pay higher prices for food, these kinds of programs transfer income from the worse-off to the better-off in American society. The practice of subsidizing tobacco and sugar, commodities that have well-known deleterious effects on human health, is also problematic.

Food Assistance

One area in which federal policies have provided support to consumers is through various types of food programs for the poor. One such program, the Food Stamp Program, has been the primary source of food assistance for low-income people. Historically, the food stamp program has been the only form of assistance available nationwide to all households based only on financial need, regardless of family type, age, or disability (Ohls and Beebout 1993).[1] This program has been consistently demonstrated to be effective in reducing hunger and improving the nutrition of people in low-income households. However, this program was severely cut in the late 1990s. Fully half of the projected budget savings from the 1996 welfare bill came from reduced expenditures in the food stamp program (Center on Budget and Policy Priorities 1996). These changes in food programs for the poor represent the largest cutbacks since these programs were first established, and will affect the nation's most vulnerable people—children, the elderly, and the disabled. This is harsh, but not surprising.

Early food programs were established not primarily to feed hungry people but to dispose of commodity surpluses purchased by the government in order to support farm incomes (Lipsky and Thibodeau 1990). In 1959 True D. Morse, the undersecretary of agriculture, said, "We are most sympathetic to the plight of needy persons. We must, however, not lose sight of the fact that the primary responsibility of the Department is to carry out the farm programs that benefit farmers" (Kotz 1969). These same kinds of

---

1. Many of the reductions in the 1996 bill came from reducing the eligibility of two groups of people: able-bodied adults with no dependents and noncitizens.

trade-offs affected funding for food stamps in 1996. For example, while Senator Richard Lugar at first opposed cuts in food stamps and nutrition programs, after heavy resistance from agricultural lobbies he produced a plan that cut nutrition programs even more deeply than budget resolution recommendations (Hosansky 1995). This new plan left more funding available for farm programs like crop subsidies and conservation incentives. There is historical precedent for this approach. At different times throughout the twentieth century hunger alleviation funds have been withheld when it was thought that they would disrupt agricultural markets (Andrews and Clancy 1993). Farm programs have always been privileged over food programs. As DeLorme and others (1992: 430) put it, "The needy, who really required assistance, were largely pawns in the rent-seeking games of others."

Economic Democracy

There are also enormous differences in the amount of funds distributed to individuals in food programs as compared to farm programs. The Food Stamp Program is targeted at low-income people, and in order to receive food stamps recipients must qualify through a strict means test. Farm programs have never been means tested. In 1949 the secretary of agriculture, Charles Franklin Brannan, introduced the Brannan Plan, which would have shifted from a price standard to an income standard as the basis for determining a "fair" return to farmers and limited government agricultural supports to a certain volume of production. The Plan was opposed by all major farm organizations except the National Farmers Union, however, and therefore was never enacted (Tweeten 1979). While food stamps make a large difference for those who receive them, they do not constitute a large subsidy to a low-income person. Food stamps are estimated to increase the recipient's food purchasing power by about 25 percent. Farm programs, on the other hand, comprise a substantial portion of farm income—50 percent in 2000 (Egan 2000). The average food stamp benefit per person in March 1997 was $72 (USDA 1997b) or $864 per year. Farm programs pay out much more than this to their recipients. For example, in 1996–97, annual payment limitations per farm ranged from $80,000 to $230,000. A provision in the 2002 farm bill to make these programs fairer by capping payment limits was defeated by large margins in both the House and the Senate. Farm programs also contribute to income inequities between the category of farmers and the category of consumers. For example, in 1986 government payments were sufficient to pay the equivalent of $42,000 to each commercial farm, a sum that is roughly $14,000 greater than the U.S. median family income (U.S. Office of the President 1987). The average farm household in 1999 had an income of $64,347, whereas the average for nonfarm household was $55,842 (Economic Research Service 2001). On the other measure of economic well-being—wealth—farmers are considerably better off than nonfarmers. In 1997, assets per nonfarm household averaged about $375,000, while the value of assets was almost twice as much for farm households, at about $700,000 (Hallberg 2001). Of course, like all aggregate data, these averages conceal uneven income and wealth distributions both among consumers and among farmers. Some are very poor, some are extremely wealthy.

In contrast to food programs that benefit low-income people, farm programs have tended to benefit the relatively higher-income farmers. Since farm support programs link benefits to acreage historically under production, the primary beneficiaries of these programs are large landowners. This occurs both because payments are based upon production levels and because they are based upon average costs (large farms with lower costs, therefore, are paid excess profits). The Freedom to Farm Act of 1996—passed the same year as the welfare reform that cut food programs to the poor—distributed most of its payments to the largest farms. The top 10 percent of recipients received 61 percent of the payments (Williams-Derry and Cook 2000). This asymmetric distribution of program payments has additional consequences. Most of the value of these programs accrues to landowners through rising rental rates and through capitalization in to land values. The operators farming mostly rented acreage (42 percent of farmers) receive little benefit from the programs (USDA 2001).

Thus, farm programs contribute to the inequality in income and wealth in the agricultural sector. Not only do farm programs disproportionately benefit wealthier farmers, they contain no provisions for the impoverished farmworkers who provide a significant share of the labor in agricultural production (Slesinger and Pfeffer 1992). Further, the USDA has historically discriminated against people of color, causing loss of land and livelihood. The "40 acres and a mule" promised to freed slaves after the Civil War, for example, never came to pass, as Andrew Johnson decided to give the land back to the previous owners. In 1997 African-American farmers filed a class-action suit representing eighteen thousand farmers against the USDA (Pigford v. Glickman), charging that the USDA has discriminated against farmers of color, denying them loans and other benefits because of their race. It was settled in an out-of-court agreement that the USDA would review claims of individual African-American farmers and make payments where the claims were upheld.[2] As of July 2003, $700 million dollars has been paid to claimants, and $18 million of debt has been discharged (USDA 2003).

There is a growing perception that U.S. agricultural policies have been ineffective in meeting their objectives. Mainstream agricultural economists made this observation twenty years ago. According to Gardner (1983: 219), "There has emerged a quite remarkable consensus of the left, center, and right that governmental intervention in agricultural commodity markets has had undesirable results in almost every instance, in every country." Little had changed in priorities of American food and agricultural policies in the twenty years since this observation was made until alternative agrifood movements became active at the federal policy level. Alternative agrifood advocates have their work cut out for them. The fact that American food and agricultural policies have enabled or failed to correct many of the environmental and human problems addressed by alternative agrifood movements is closely related to the undemocratic policy environment from which they have emerged.

Inertia and Power in the Policymaking Process

Alternative agrifood movements will not be able to change American food and agricultural policy unless they can somehow overcome the characteristic inertia of the federal government in this policy area and somehow overcome or outmaneuver the structures of power and privilege that originally created and continue to maintain these policies. The goals and mechanisms of U.S. agricultural policy have changed little since the New Deal; once policies are adopted they may be supplemented but are rarely altered in any fundamental way. The "laws of motion" of U.S. agricultural policy are such that major policy changes are rare and are developed in response to crisis; once adopted, these policies tend to persist (Robinson 1989). Most policy debates are formulated in extremely narrow terms, and suggestions for significant policy changes are vaporized in the committee negotiation process. As Browne and Cigler (1990) phrase it, "Today's agriculture groups are not interested in remaking the political world, as were their many predecessors in the last century."

---

2. Many plaintiffs in the case have been dissatisfied with the process and asked to initiate another lawsuit against the USDA, but were denied.

A further obstacle to policy change is the material and cultural relations of power in the U.S. agrifood system that have configured agricultural policy. Agricultural policies create significant benefits for certain groups of individuals, who, therefore, organize politically to maintain and maximize their benefits. Throughout history agricultural policy has been made by growers, nonfarm agribusiness interests, the USDA, and the land-grant universities. One measure of the influence of the food and agricultural industry is its financial prominence in Congress. For example, agricultural interests contributed $24.9 million to presidential and congressional candidates in the 1991–92 elections, averaging $76,000 to each House agriculture committee member and $123,000 to each Senate agriculture committee member (Makinson and Goldstein 1994). These contributions were fifty times greater than those for either health and welfare or children's rights.

Most large businesses and organizations have at least one political action committee (PAC), the purpose of which is to channel resources to legislators. PACs are quite prevalent and active in the agricultural sector. Some agricultural PACs, such as those related to dairy, rank near the top of all U.S. Political Action Committees (Knutson et al. 1998). Nearly all food and agriculture committee contributions come from producers, business, and industry, with a very small amount coming from consumer or labor groups. An analysis that examined this distribution in the 1970s found that of the food and agriculture PAC contributions to candidates for Congress in 1977–78, 61 percent came from producers, 33 percent from business and industry, and 6 percent from citizen, consumer, labor, or professional groups (Guither 1980). More recently, the Environmental Working Group (EWG) tracked Food Chain Coalition Political Action Committee campaign contributions. The Food Chain Coalition is comprised of over 230 corporations, trade associations, and organizations representing farmers, pesticide manufacturers, farm suppliers, food processors, and retailers. The EWG found that the top five contributors with PACs from the Food Chain Coalition in the 1994 and 1996 election cycles were RJR Nabisco, Phillip Morris, American Crystal Sugar, PepsiCo, and the National Cattlemen's Association (Davies et al. n.d.). The agrifood system we have today is the product of the power relations that have shaped the organization and practice of agriculture and reinscribed these power relations in political institutions.

Legislative decision makers have represented a narrow spectrum of interests belonging to the most powerful in our society. Decision making has been asymmetric not only along economic lines, as seen in financial con-tributions to Congress, but also along dimensions of race and gender. According to Dunn (1993), only 11 percent of those in the House of Representatives and 7 percent of Senators were women, most of them recently elected and thus the most junior and least powerful members of Congress. This disproportionate representation is magnified in agricultural committees, where hardly any women, people of color, or low-income peo-ple have served. Out of the fifty members of the House Committee on Agriculture of the 107th Congress, there was one woman and four minor-ity members (House n.d.). Similarly, out of twenty-one members of the Senate Agriculture, Nutrition and Forestry Committee, there were two women members and no minority members (Senate n.d.). Most of the members on these agricultural committees are from farm states that ben-efit directly from farm programs. Historically, U.S. agricultural policy-makers, business and farm-group leaders, researchers, and educators have been predominantly affluent European-American men. Women and peo-ple of color, who do much of the work in agriculture and who represent a disproportionate percentage of the poor, have been conspicuously absent from key agricultural decision-making positions. Agricultural policy deci-sions are made in these small committees, resulting in a nearly complete agribusiness "lock" on Congress (Merrigan 1997). Merrigan has observed that this traditional power structure—the "iron triangle" of agricultural interest groups, agricultural legislative committees, and the administrative agencies of the USDA—has excluded broad-based public debate on agrifood issues. Furthermore, Merrigan reports that when food and agriculture issues become public, those who are called upon to testify before Congress are the well-financed interest groups rather than a broad spectrum of those affected by the issues.

Increasingly, however, there have been fissures in the iron triangle as more and more nontraditional groups have gotten involved in food and agricultural policymaking. Although conventional agricultural interests—a small group with few differences in viewpoints—have always dominated the agricultural policymaking agenda, in the late 1970s a "New Agenda" began to be articulated by groups who had not previously been involved in the agricultural legislative process (Paarlberg 1980). Around this same time, the assumptions of New Deal agricultural policies began to disintegrate (Swanson 1989). Both within and outside agriculture, serious dissatisfac-tion with agricultural policies grew, creating a demand for a "bold reori-entation" of current policy (Rausser and Farrell 1985). The confluence of these factors has resulted in some changes in the policymaking process.

Participation in discussions and decision making (although not at the highest levels of power) has been opened up to nontraditional interest groups. The expansion of these forums represents a significant improvement over traditional decision-making arrangements in food and agricultural policy. Through the work of alternative agrifood movements and the institutionalization of sustainable agriculture and community food security programs, the long-standing antagonism and divisions between traditional agrifood institutions and dissenting groups has been reduced. What will be required to maintain these programs and make further progress toward sustainability and sustenance in food and agricultural policies?

The fact remains that politics responds to two basic forces: money and constituents. It is unlikely that alternative agrifood groups can muster the raw financial power to change policy, at least in the short term. Although "new agenda" groups have begun to enter agricultural policy debates, they are far outspent and outpowered by traditional agricultural interests. For example, the annual budget of one large commodity organization exceeds the combined annual budgets of the top ten organizations promoting sustainable agriculture (Merrigan 1997). In a policy process that is never-ending and occurs in many different places, Merrigan does not think it is possible for sustainable agriculture compete with well-funded agribusiness organizations.

Another type of "expense" in the policymaking process is that of transaction costs, that is to say, the costs of participating in the negotiating process, from gathering and disseminating information to participating in social activities. Those that have been best organized politically to shape agricultural policy have been those with vested economic interests in the outcomes of these policies, primarily agricultural producers, agricultural input suppliers, and commodities traders. Since these firms are financially dependent upon federal agricultural policies, they allocate time and money to policy formation as a cost of doing business. By contrast, the impacts on the nonagricultural groups emerging on the agricultural policy scene are generally too small to warrant their spending resources on organizing around a single agricultural policy issue.

Consumers worried about food safety, environmentalists concerned with the degradation of soil and water, and social reformers advocating for the hungry and the poor—have all historically had extremely high costs for their participation in agricultural policy. Because these concerns are peripheral to how these consumers, environmentalists, and social reformers make their livings, their groups are few, small, poorly funded, and unable to lobby on behalf of more than a few policies per year. Although both the CFSC and the National Campaign for Sustainable Agriculture each have a lobbyist in Washington, D.C., they cannot deal with every piece of food and agricultural legislation that comes before Congress, as can the hundreds of agribusiness lobbyists. Merrigan (1993) points out that advocates of sustainable agriculture had to work so hard to create sustainable agriculture legislation in the 1990 farm bill, that after the bill became law, many advocates were too tired and underfunded to pursue implementation and resource allocation. While the conventional agricultural system has been produced by a coterie of the powerful and privileged, alternative agrifood movements will need to build coalitions in order to gain power. Developing a politically powerful coalition with enough clout to change long-standing food and agricultural policies will mean forming new alliances, both conceptual and strategic. This can happen both at the level of discourse and through practical alliances.

Importance of Discursive Clarity

A collection of groups that have similar positions on issues can be characterized as a discourse coalition or discursive alliance. A discourse coalition is a group of actors that share and maintain a particular way of thinking and talking about an issue, that is, they share "story lines" (Hajer 1995). Story lines are narratives of social reality that provide actors with a set of symbolic references that suggest a common understanding of their situations and social conditions. They are essential devices that allow participants to overcome social fragmentation and achieve discursive closure on issues of common interest. The actors in these discourse coalitions do not necessarily meet each other and do not necessarily have an articulated and agreed upon strategy, but they share perspectives on central questions of common concern. For example, the environmental justice and community food security movements are separate, yet they have parallel goals and intersecting agendas (Gottlieb and Fisher 1996a). While alternative agrifood movements in and of themselves constitute a type of discourse coalition, their effectiveness would be increased through the development of a clear conceptual statement of priorities and positions.

One could argue that the concepts of sustainable agriculture and community food security have become prevalent because they are interpretable in enough different ways to satisfy constituents from widely diverse perspectives. In a discourse-coalition approach, the effectiveness of the discourse derives from the fact that it can be interpreted in many different ways (Hajer 1995). Of course, such flexibility can be the undoing of a discourse coalition if the tenets and concepts around which it has organized are too malleable or ambiguous.

Vague definitions also make them vulnerable to co-optation or misinterpretation. For example, Youngberg and others (1993: 300) point out that the symbolic power of sustainability is such that fertilizer, pesticide, and genetic engineering technologies all now "reside under the sacred temple of sustainability." Most of these technologies would be excluded from—and are perhaps antithetical to—the ideas promoted by most advocates of sustainable agriculture. For example, an atomic energy journal even has an article on "Atoms for Sustainable Agriculture: Enriching the Farmer's Field," which looked at how nuclear and isotope techniques could be used to improve soils and sustain crop production (Hera 1995). Agrochemical companies such as Bayer, Syngenta, BASF, and Dow have adopted the term "sustainable agriculture," pointing out that their chemical crop-protection technologies are key to achieving "optimal yields" in sustainable agriculture. There can be a downside to discursive and institutional success.

Clear definitions are also crucial for engaging people and bringing them together to work on common objectives. Bob Gottlieb, cofounder of the community food security movement and a former board member of the CFSC, argues that a conceptual big picture is necessary to bring together the diverse and innovative projects and programs operating under the umbrella term "community food security." Gottlieb opines that the term remains vague to activists and policymakers and even to those in the movement itself. He summarizes, saying that "community food security has to be more than just a poorly defined term that confuses even those who identify with it" (Gottlieb 2003: 7). Merrigan (1993) suggests that the lack of a clear representation of what sustainable agriculture actually means serves to restrict public involvement in legislative debates on sustainable agriculture. While loose definitions allow for coalition building in certain instances, ultimately they can restrict progress toward movement objectives.

Despite the potential for co-optation and missed opportunities for galvanizing a constituency, many advocates of sustainable agriculture are little interested in further articulating the meaning and indicators of sustainability. Many publications about sustainability or sustainable agriculture assert that the term "sustainable" is ambiguous and undefined, and is probably best left that way. In part, this is a strategic position for increasing the acceptance of the term. Advocates of sustainable agriculture have been "lured to a definition that would offend no one and promise something for everyone" (Youngberg et al. 1993: 310). It has become almost fashionable in sustainable agriculture circles to ridicule the idea of trying to define sustainability. In the California survey, some respondents from organizations promoting or advocating sustainable agriculture expressed exhaustion with the efforts to come to a common definition. One respondent, instead of stating their definition, wrote, "Hasn't this been done over and over in the last twenty years?" Now that the term has been widely embraced, there has been a tendency to want to close off debate on its meaning and get on with the project of making agriculture sustainable.

While spending time articulating and refining definitions of sustainable agriculture and community food security may seem like an academic extravagance, there are important conceptual and practical reasons for creating clear definitions. As Lockeretz (1989) wrote more than a decade ago, while everyone seemed to be getting on the sustainability bandwagon, given that its most basic ideas remained to be worked out, "Isn't something backwards here?" A Kellogg IFS Initiative study concluded that "sustainable pioneers require a clear picture of where they're heading to know if they're moving in the right direction" (Scheie 1997: 10). Or, as the Cheshire Cat asks in *Alice in Wonderland*, if you don't know where you're going, how will you know when you get there? If there are no goals, how do advocates, researchers, and practitioners know when they are working with each other or against each other, or if they are getting closer to or further from their goals?

Connections Within Alternative Agrifood Movements

Alternative agrifood movements increasingly see problems and solutions in the agrifood system as interrelated, spanning the range from preproduction to postconsumption. For example, many of the groups and institutions most interested in community food security are those with strong interests in sustainable agriculture (Gottlieb and Joseph 1997: 10). And correspondingly, the community food security movement supports family farms and organic agriculture. The U.S. Action Plan on Food Security includes a section on sustainable agriculture (USDA 1998a). This is also happening at an international level. In 1996 an international conference titled "Globalization, Food Security, and Sustainable Agriculture" focused on how to create improved food systems around the world (Food security internationally 1997). These connections were prefigured by the Cornucopia Project (1981), which brought together environmental issues with concerns over food safety and food availability back in the early 1980s.

A number of earlier agrifood alliances have focused on the big picture rather than single issues. For example, in the United States, when the Farmer's Alliance ran into commercial resistance to its cooperatives, it became explicitly political with a broad agenda (Adamson and Borgos 1984). Their platform included regulation of the railroads, expansion of the national money supply (to lower interest rates), legal recognition of trade unions, and a tax on speculative real estate profits.[3] In Canada, instead of focusing only on aspects of the economy that could be controlled within the province, farmers went on the offensive against the total economic structure (Lipset 1968). The agenda was based upon a political economy that would provide for human needs rather than profits and included priorities such as the socialization of health care, changes in labor law, and social ownership of productive resources (Winson 1993). These kinds of efforts recognized systemic problems and common interests among diverse groups.

Increasingly, conceptual connections are being made between the two main sections of alternative agrifood movements: sustainable agriculture and community food security. Although the sustainable agriculture movement has developed with environmental and populist interests in the lead, it is beginning to prioritize ideas of community food security and local food systems. As one example, the National Consortium for Sustainable Agriculture Research and Education renamed its newsletter to include food in the title; it is now called *Inquiry in Action: Learning Partnerships for Sustainable Agriculture and Food Communities*. The National Campaign for Sustainable Agriculture's policy priorities include both environmental and social justice priorities. For example, it advocates policies to provide increased opportunities and support for ethnic minority farmers, as well as policies for farm-based conservation programs, and community food security is a prominent part of its platform. The California Sustainable Agriculture Working Group (CalSAWG) has produced a white paper on social concerns in sustainable agriculture (Inouye and Warner 2001) and in

---

3. As the Alliance expanded and became increasingly political, it started a third party, the People's Party, which ran state, local, and presidential candidates, a strategy that was eventually broken by the Democratic party.

early 2003 addressed issues of social justice at its annual board of directors meeting. CalSAWG has taken up farm labor issues as one of its three priority areas.

These connections are practical as well as conceptual. For example, the governing boards of the CFSC and the National Campaign for Sustainable Agriculture have members in common. In California, the Sustainable Agriculture Working Group has worked closely on organizing workshops and other projects with the CFSC, for example, by cosponsoring the first Community Food Security Summit in California. Given that it is unlikely that alternative agrifood groups will soon have the resources to compete for political clout with major agribusiness firms, it is important that they are beginning to work together.

Alternative agrifood movements are joining together as coalitions to maximize the effectiveness of scarce resources. Starting with the 1981 farm bill, three different sets of advocates—environmentalists, proponents of sustainable and organic agriculture, and advocates for the small family farm—joined forces to seek changes in federal food and agriculture legislation (Gottlieb 2001). These groups worked on the 1985 and 1990 farm bills and made efforts to connect antihunger groups to the alternative agrifood agenda. For example, the 1985 Food Security Act addressed environmental issues in much greater measure than any previous farm bill, and sustainability figured prominently in the debates on the 1990 farm bill.[4] For the first time, the 1996 farm bill brought together urban food interests, advocates for sustainable agriculture, farmland preservation groups, and advocates for rural development. This kind of coalition building is especially important in the face of highly entrenched agricultural interests in national policymaking. The USDA SARE and Community Food Projects programs would never have been established if these new groups had not entered the agricultural policy debates. These examples show an increasing coalescence of rural and urban, consumer and producer, and environmental and social concerns in alternative agrifood movements.

Another illustration of how combining the efforts of alternative agrifood movements can have significant impact on agrifood policy is the outcome of the fight over the National Organic Rule. The National Campaign for Sustainable Agriculture and the CFSC joined forces with organic food

---

4. The two major periods in which agricultural conservation has become a political issue—the 1930s and the 1980s—were both times in which surpluses caused farm prices to drop, leading to economic crisis for many farmers.

interests to defeat the proposed USDA national standards for organic produce, which in their view would have compromised the quality and reliability of organic food. Working together, these groups orchestrated initial opposition to the proposed federal organic rule and catalyzed public interest in further participation. As a result of these efforts, the National Organic Standards Board solicited and received more public input on the organic standards than any previous rule (Guthman 1998). Despite this input, however, the rule eventually proposed by the USDA differed substantially from the organic practices endorsed by organic farmers. Alternative agrifood movements were particularly concerned with the intent on the part of the USDA to allow the use of genetically modified organisms, sewage sludge in organic production, and food irradiation.

The release of this version of the rule galvanized an even higher level of citizen participation than before. More than a thousand newspaper and magazine articles and radio and television broadcasts appeared in the six months after the rule was made public; news of the rule made the front page of the *New York Times* twice in six months (Bowen 1998). During the four-month public comment period, the USDA received more than 220,000 comments on the rule from producers, consumers, environmentalists, and others—the largest public response the USDA had received to any proposal in the memories of the people at the USDA (McCann 1998).[5] The intensive, combined organizing effort on the part of alternative agrifood groups led the USDA to retract the proposed rule and go back to the drawing board, ultimately resulting in a rule that could be endorsed by most organic farmers.

This kind of success at the federal policy level has continued. The 2002 Farm Act contained several first-time provisions to support organic agriculture. For example, it authorizes $15 million in new funding for organic research and $5 million for a cost-share program to help small-scale growers with the costs of federal organic certification. The same legislation doubled the budget for the USDA Community Food Projects program. At the same time, there were also significant movements toward greater inequity in the 2002 farm bill. Despite House and Senate bills that would have increased conservation funding and reduced crop subsidies, ultimately, crop subsidies were tripled, payment limitations were not enacted, and conservation was funded at lower levels than originally proposed in the bills. The

---

5. However, much of the organizing around the organic rule was through the internet, a medium which is less accessible to low-income people.

CFSC is looking for opportunities for policy change outside the traditional agricultural legislative process. For example, the Coalition advocates funding transportation programs to increase food access through the Federal Transportation Equity Act (Gottlieb 2003). Another example is working to develop farm-to-cafeteria projects through the House Subcommittee on Education Reform. The Coalition has been developing connections and alliances with other groups to increase their ability to influence these major pieces of legislation. Sustaining and increasing the political power of alternative agrifood movements will require this kind of continued alliances among different constituencies.

There is ample precedent in agrifood movements for joining constituencies that seem to have little in common, or who may even be opposed to each other. Although food and farm issues have tended to mobilize different constituencies, at times their connections have been made explicit. For example, Albert Howard's (1943) study and advocacy of organic production methods was inspired by his concerns about hunger and long-term food security. Some of the early success of the UFW was due to its ability to organize for social justice among both field workers and urban consumers. Farmers and urban labor were allied in the early 1930s as part of the Co-operative Commonwealth Federation in Canada. Lobao and Thomas (1992) have found that farm operators who supported progressive agendas in the farm sector (e.g., support for agricultural research for smaller farms and increased taxes for corporate farms) also supported redistribution of public resources toward the poor and working class. Of course, expanding and deepening these kinds of conceptual and strategic alliances will not be simple or straightforward.

Challenges of Connecting Agrifood Constituencies

Finding sufficient overlap in interests and priorities among the various groups involved in the agrifood system has been a challenge. For example, family farmers are more likely to support, say, "right to farm" laws than they are to support environmental regulations or public programs for the hungry (Magdoff et al. 1998). As Mooney and Majka (1995) point out, although both farmers and farmworker movements are grounded in agricultural production, they have divergent class interests and vastly different access to the power and resources needed to realize those interests. This tension between farmers and farmworkers is one with which the sustainable agriculture movement continues to struggle. Organizations that have raised the question of social justice for farmworkers have historically encountered resistance from even small or organic farmers and their organizations. This is possibly related to the economic dependence of California farmers on hired farmworkers, who are almost always different from the farmers in ethnicity, class, and citizenship.

While farmers support government payments to farmers, they have generally been opposed to government assistance for low-income people (Clancy 1993). Wimberley (1993: 1) illustrates several other possible points of divergence based on social location:

> Consumers may not agree with agricultural interests and agricultural interests may differ from rural nonfarm interests. Urban citizens may neither know nor care about rural conditions. Agricultural producers may feel that urban people should have little to say in farm matters. Animal rights activists confront ranchers and packers. Farmers argue with environmentalists. Food-safety-conscious consumers challenge biotechnology and chemicals used by farmers and processors. Poor families want food at low cost. Agricultural wastes offend nonfarm rural residents. Rural residents wonder why their infrastructure, services, and employment lag so far behind that of urban areas.

In food and agriculture, environmental and social justice issues may sometimes be at cross-purposes. For example, because of concerns about the environment, older pesticides have been replaced by those that are less persistent in the environment and food. It turns out, however, that these new pesticides are more acutely toxic for workers (Wasserstrom and Wiles 1985). One of these pesticides, parathion, is used today in place of DDT because it breaks down more quickly in the environment. The problem for workers is that it releases a chemical that is fifty-five times more toxic than the parent chemical when absorbed by human skin. While producer and consumer groups do share common interests, it would be inaccurate to suggest that they have the *same* interests.

While the problems addressed by alternative agrifood movements do connect social and environmental issues, these connections are fairly recent. For example, despite efforts to connect production and consumption, until recently, access to food—a basic social justice issue—was not on the agenda of the American sustainable agriculture movement. Many advocates of sustainable agriculture saw no connection between sustainable agriculture and hunger (Clancy 1993). The results of the CASA 1996 California questionnaire, for example, showed that out of fifteen issues, sustainable agriculture groups thought hunger and food access were the least important. In contrast, the International Movement for Ecological Agriculture's (1991) declaration on ecological agriculture begins with the issue of hunger. In the United States, attention to food-consumption issues had focused primarily on toxin-free food rather than food security. An illustration of this perspective is Dicks's (1992) statement that the abundance of the U.S. food supply means that sustainability in the United States need not be about hunger or poverty, which are problems only in "socialist" and "Third World" nations.

This tendency to separate environmental and social justice issues goes back to the beginning of the sustainable agriculture movement. The movement for sustainable agriculture has aspects of both environmentalism and social justice (Buttel and Gillespie 1988), but the social justice priorities diminished in importance over time. The environmentalist origins of the movement are represented by groups such as the Rodale organization, which promotes a style of organic farming originally formulated by European thinkers. Rodale's vision includes small-scale production, local marketing, low mechanization, and the absence of artificial soil amendments and biocides. This vision was articulated within a framework of agrarian fundamentalism, a critique of industrial society, and appeals to personal health and nutrition. The social justice aspect of the movement, represented by groups such as the New World Agriculture Group and the Rural Advancement Fund, concentrates on issues such as farmworker struggles and developing a resource-conserving, noncorporate food and agriculture system. Their positions were often based on a critique of advanced industrial capitalism in agriculture. In some instances, sustainable agriculture groups tended to focus more on impoverished countries than on the United States because they believed that social injustice was less pressing here. Ultimately, the environmental and populist aspects of the sustainable agriculture movement took precedence over the social justice aspects in the United States.

Shades of these differences can be seen in the alternative agrifood movement's actions around the 1996 farm bill. While more alternative agrifood groups joined together than ever before, this process also resulted in the emergence of disagreements between progressive farming factions and representatives of the urban poor (Gottlieb and Fisher 1996b). That these groups had different perspectives and priorities was also illustrated in the findings of our 1996 California survey of alternative agrifood organizations. In this study, the widest divergence in perspectives from those of the sustainable agriculture groups was found not in conventional agricultural groups, but in the food-issue groups. Food-issue groups were much more concerned with concentration in ownership, hunger and food access, variations in distribution of income in food and agriculture, public education priorities, and diversity in agricultural decision making than sustainable agriculture groups were.

Since this time, alternative agrifood movements and institutions have begun to show an interest in social justice issues such as food security, although environmental priorities still tend to dominate their agendas. In our recent study of California alternative agrifood institutions, 80 percent of the respondent organizations were practicing or advocating environmentally sensitive production methods (Allen et al. 2003). Of the other 20 percent of organizations, half said they supported sustainability, just not at the expense of making food more affordable or making it so farmers would not be able to earn a decent living. For these organizations, sustainability is a priority, but it is not subordinate to the priorities of social or economic justice. This position was held by organizations working primarily with a low-income or small-farm clientele. Only 10 percent of leaders said their organization did not have a position on environmental soundness, or that there was a "diversity of perspectives" in their organization. The majority of organizations that responded in this way were those oriented toward marketing agricultural products. None of the organizations in the sample were hostile to the vision of an environmentally sound food system. This represents very strong support for moving toward a more sustainable agrifood system, even where the organizations were not themselves directly focused on sustainability. The emphasis on social justice was weaker, however, with only a third of the organizations focused on this as a priority. In contrast, half of the California alternative agrifood organizations that responded to a similar question on social justice in the agrifood system had no position at all on social justice, and a few were hostile to the concept. Furthermore, what people meant by social justice was quite disparate, ranging from changing basic political-economic structures to repeating priorities of environmental soundness.

Alliances for Social and Environmental Justice

The practical and conceptual tensions between environmentalism and social justice are by no means unique to the agrifood system. Since the rise of the environmental social movements of the 1970s, there has been friction between social justice and environmental priorities within the movements, strains that have not always yielded to attempts at reconciliation (Pepper 1993). Atkinson (1992) suggests that a key difference that leads to these tensions is that at the root of environmental problems environmentalists see industrialization, while social justice advocates see capitalism. Increasingly, however, both conceptual and strategic connections are being made between environmental and social goals in social movements. For example, alternative agrifood movements joined with coalitions of labor, women, and environmentalists in the fight against the North American Free Trade Agreement (NAFTA). On a more local level, Gottlieb and Fisher (1998) report that environmental justice and food security objectives have come together in projects in Los Angeles.

The argument for making connections between social and environmental objectives and among scales is illustrated by the case of the anti-toxics movement, spawned by the hazardous waste catastrophes of the late 1970s, such as Love Canal. Although many anti-toxics groups began with a "not-in-my-backyard" (NIMBY) orientation, the movement quickly confronted basic class, race, and gender issues and became a "widespread, dynamic social movement" at a time when other social justice movements were on hold or in decline (Szasz 1994). Membership in the movement also played a key role in that many of the organizers had roots in the social justice movements of the 1960s, such as the civil rights movement and farmworker unionization (Di Chiro 1992). As the anti-toxics movement pushed beyond personal human health issues and began to universalize rights to health for everyone, it entered the terrain of modern, industrial production and asymmetric distributions of power. This change was almost foreordained because, as Szasz makes clear, the high correlation between toxic exposure and poverty forced the examination of overall economic conditions and systems. The same correlation between toxic exposure and communities inhabited by people of color required the examination of racism. Finally, the fact that the movement dealt with traditional women's terrains of family and health, combined with the fact that most of the activists were women, required a confrontation with patriarchal modes of domination. Through these observations and experiences, the anti-toxics movement came to embrace an environmental justice framework and focus simultaneously on issues of class, race, and gender.

An example of an agrifood movement that connected environment and social justice is the effort to organize farmworkers in California's strawberry industry in the late 1990s. The UFW targeted strawberries as a commodity in which to organize farmworkers in 1996. To win public support for the workers' struggle, the UFW and the AFL-CIO organized a national campaign called "5¢ for Fairness." The campaign was based on an analysis that showed that a five-cent increase in the price of a pint of strawberries could result in a 50 percent increase in the wages of most workers. The premise was that since the strawberry industry already taxes itself for research and marketing programs, it could afford to increase farmworker wages through a similar mechanism. Precisely because strawberries are a luxury food, marginal price increases are likely to be tolerated by consumers. Environmentalist support for the Campaign came from the fact that strawberries are the most pesticide-intensive crop in California. A large proportion of the pesticide use is the agricultural fumigant methyl bromide, a Class 1 ozone-depleting substance. There are no readily available substitutes for methyl bromide, meaning that the planned 2005 phaseout of the product could require changes in the agroecological and social organization of strawberry production.

Another point of convergence is that pesticides pose common—although unequal—risks to consumers and farmworkers. For example, the destruction of the ozone layer, caused in part by the use of methyl bromide, poses a risk to all people, irrespective of class, gender, or ethnicity, and is therefore a common issue for organizing. At the same time, methyl bromide is a class issue in that farmworkers who apply the compound are at much greater health risk than are those who are not farmworkers. The strawberry campaign illustrates possibilities for uniting environmental and social justice issues into a coherent social movement.

The strawberry campaign represented the largest bottom-up labor organizing drive in the country at the time. It established the first coordinating body among more than six hundred central labor councils around the country, which took the effort into the supermarkets. This campaign was looked to as a new style of labor organizing that could inspire similar progressive efforts in other industries (Bacon 1997). More than forty organizations—including traditional labor, civil rights, religious and environmental groups—joined in this effort. These included the NAACP, the

National Organization for Women, and the Sierra Club. In addition to traditional labor issues, such organizations represent struggles against racism, environmental destruction, and unsafe food. Although the efforts to unionize strawberry workers had limited success, the strawberry campaign was one of the first concrete instances of a social justice and environmental coalition in food and agriculture in the United States.

Efforts that unite social and environmental justice in food and agriculture are increasingly taking shape at the international level as well. The Rio Earth Summit in 1992 and the World Food Summit in 1996 prompted diverse groups from around the world to seek changes in global policies and practices. These groups, which include farmers, peasants, trade unions, community organizations, and indigenous peoples, have come together around the issues of sustainable agriculture and food security (Elswick and Forster 2002). Key issues include access to resources for food production, establishing food as a human right, improving working conditions and wages in the agrifood system, changing rules around intellectual and genetic property rights, and changing trade rules that disadvantage small-scale agriculture. In September 2001 representatives of sixty countries met in Havana, Cuba, to discuss these issues at the World Forum on Food Sovereignty. In preparation for the 2002 World Food Summit, these organizations assessed the little progress that had been made on eliminating hunger since the World Food Summit in 1996. They propose three themes as key elements in strategies to end hunger and malnutrition: (1) the need to develop a rights-based approach to hunger and malnutrition issues, (2) the need to limit corporate control of the food system that has negatively affected small farmers and indigenous communities, and (3) the need to promote agroecological models of production (International Planning Committee for the World Food Summit 2002). These principles are being translated into practice through a variety of efforts.

One effort, Social Accountability in Sustainable Agriculture (SASA), is a collaborative project of four international labeling organizations: Fair Trade Labeling Organizations (FLO), Sustainable Agriculture Network (SAN), International Federation of Organic Agriculture Movements (IFOAM), and Social Accountability International (SAI). The goal of the project is to develop guidelines and tools for the implementation of audits of social standards in sustainable agriculture. The basic standards of a major international organic certifying body, the International Federation of Organic Agriculture Movements (IFOAM), has included social justice in its list of principles since 1996. Its goal is to "progress toward an entire production, processing, and distribution chain which is both socially just and environmentally responsible." Further defining and operationalizing this goal is part of IFOAM's work plan, hence its participation in the SASA project. In August 2002 an international group of people met before the biannual IFOAM conference to discuss how to proceed and provide input into a draft document, "Toward Social Justice and Economic Equity in the Food System," written by Elizabeth Henderson, Richard Mandelbaum, Oscar Mendieta, and Michael Sligh—a group that has also been working for several years on developing social justice standards for sustainable agriculture in the United States.

These efforts build upon the international fair-trade approach that offers criteria for an alternative model of development and trade based on participatory democracy, sustainable development, social justice, and respect for cultural and ethnic diversity. What started as an aid to European recovery in postwar Europe led to the creation of fair trade core business values in the 1970 and 1980s as this European model was posited as a way of improving the lives of low-income producers in impoverished nations. As an alternative to free trade, fair trade works to create means and opportunities for producers—especially disadvantaged, small-scale producers—to improve their living and working conditions. The overall mission of fair trade is to promote equity, environmental protection, and economic security through changing the terms of trade and expanding public knowledge about these issues. Both products and companies can be labeled and certified as being Fair Trade. In order to be certified, they must comply with the seven principles of fair trade: fair wages, a cooperative workplace, consumer education, environmental sustainability, financial and technical support for those who need it, respect for cultural identity, and public accountability. Fair Trade improves the lives of many by working through the market. In the long run, however, it is unlikely that market-based solutions can comprehensively address problems that stem from fundamental social, economic, and political inequities. There is always a tension between working toward achievable goals and working toward deep social change.

9

working toward sustainability and sustenance

At the dawn of the new century people are taking charge of their food system and have given rise to a remarkable range of new discourses and practices that seek alternatives to the conventional agrifood system. They have developed a movement that simultaneously resists environmentally and socially destructive relations in the global agrifood system and works to build viable alternatives to these relations. Working toward agrifood system change involves politics at every level: personal, ideological, and institutional.

At this point, the contemporary American food and agriculture system sustains neither humans nor the environment. Agricultural policy and administrative agencies in their current forms are unlikely to develop effective solutions to problems of poverty, poor health, and environmental degradation. Instead, new agents of change are needed to develop a new united platform through "the redefinition of issues centered on the production and consumption of food" (Friedmann 1995: 25). Through their ideas and practices, alternative agrifood movements are positioned to become these agents. They have brought attention to many crucial issues of sustainability and sustenance in the American agrifood system, which has led to the creation of sustainable agriculture and community food security programs in traditional institutions. These institutions are positioned to become sites of transformation toward a food system that can sustain all of us. These groups have done an incredible job of making progress toward goals of sustainability and sustenance within the context and frame of traditional agrifood institutions. Agricultural sustainability and community food security programs are becoming institutionalized, and alternative agrifood movements are actively connecting social and environmental goals. This success at the institutional level begs the question of the role of reforms versus deeper change in the agrifood system.

Tension Between Incremental and Structural Change

The extent to which social movements should focus on incremental reform versus structural change is a central dilemma. For Raymond Williams (1973), practices, experiences, meanings, and values that are not part of the effective dominant culture can be either *alternative* or *oppositional*. Alternatives to the dominant effective culture are expressed by a person or a group finding a different way to live and wanting to be left alone with it. Oppositions are expressed when the person or group wants to change the society based on this discovery or new way of doing things. Williams has observed that small-group solutions to social problems are generally oriented toward alternatives, while large-scale movements focus on opposition to the social and political system itself.

In the context of agrifood systems, alternative meanings, values, and practices are those that do not provide a deep critique of the existing dominant culture and practice of conventional agriculture and food distribution. Instead, they offer another way of doing things, an alternative. Oppositions are based on a deep critique of conventional agriculture and food distribution and propose meanings, values, and practices that challenge and restructure the very core of the food and agriculture system. Both alternative and oppositional strategies are necessary to create short- and long-term change in the agrifood system.

Alternatives consist of different ways to accomplish established goals within the existing overarching social structure. There is wisdom in picking winnable campaigns that can build broad bases of support and energize the movement. Alternative agrifood movements have been smart and strategic in this way. One example is the Community Food Security movement's support of a law that would ban the sale of unhealthy food in California elementary schools beginning January 2004.[1] This was an issue that galva-

---

1. California SB 19 is a bill to regulate the sale of unhealthy food and beverage items in all public schools (by setting standards for portion size, fat and sugar content, etc.). After being

nized a significant number of participants at the California Food Security summit held in June 2003 because it is a concrete, targeted issue that, at least among the group at the conference, is noncontroversial. Social and political change is an iterative process, and useful actions can be taken in the short term, even if they are insufficient for long-term change. For example, while acknowledging that the most effective remedy to food insecurity is increasing the income of the poor, in the meantime Rogers (1997) suggests the creation of "nutrition stamps" that would be used for the purchase of fruits and vegetables for low-income consumers (since this is the population with the lowest consumption of fruits and vegetables).

What about oppositional perspectives and strategies? Some, such as Kloppenburg and others (1996) have claimed that "neither people nor institutions are generally willing or prepared to embrace radical change." This sentiment is evidenced by the scarcity of oppositional strategies among the California alternative agrifood institutions we studied. However, while the solutions to agrifood system problems articulated by the organization leaders were incomplete, most organization leaders were very aware of this shortcoming and not particularly content with it. For example, one interviewee said that their version of an ideal food system would be one in which land was owned in common, but they quickly pointed out that they did not believe that we would ever achieve this situation in the United States. For many of the respondents, there was a general sense that people were doing what they could, where they could, within a context of overwhelming structural impediments to an environmentally sustainable and just food system. Perhaps this accounts in part for the extent to which the development of local food systems has captured the imaginations of many organizations. In the California study, many leaders felt that the scope and depth of agrifood-system problems were beyond what their organization could address. Instead they sought out spaces where they could do something that would contribute to a better food system, however they defined it. The problem with focusing on structural change is that it seems remote and impossible.

---

amended to apply mainly to elementary schools, SB 19 was passed and signed by the governor, and scheduled to take effect in January 2004. Although the bill was modified to not regulate food sold in high schools, it contained a pilot program to provide LEAF (Linking Education, Action, and Food) grants to middle and high schools willing to adopt nutrition policies paralleling the bill's elementary school requirements. Several California school districts such as Los Angeles, San Francisco, and Oakland are basing further changes on the intent of the bill. For example, the Los Angeles Unified School District Board unanimously voted to ban soft drinks in all LAUSD schools, not just elementary schools.

It can take time and energy away from work that could be making a difference in people's lives even if it does not change the basic system. These leaders were more overwhelmed by than they were unwilling to confront core problems in the agrifood system.

Nonetheless, there is a tension between pursuing oppositional goals and gaining the institutional acceptance necessary to make incremental progress. In order to gain approval for the movement within the mainstream agrifood industry, advocates of sustainable agriculture have been "anxious to avoid confrontation with the defenders of conventional agriculture" (Youngberg et al. 1993: 310). For example, CASA was reluctant to work on projects that might alienate the agricultural community. Most CASA board members held comprehensive views of the meaning of agricultural sustainability and supported the idea of far-reaching social change. However, it turned out to be difficult to prioritize projects that would be of little interest to or contrary to the short-term goals of traditional agriculture, particularly since CASA saw building bridges with the agricultural community as an important strategy for changing the agrifood system.[2] Campbell (2001: 353) sums up the situation, describing the sustainable agriculture movement as being "caught in unyielding tradeoffs between their commitment to deeply rooted social change and the need to be politically credible." As a result, most projects and efforts remain in the realm of alternatives.

The danger of working only at the level of alternatives is that they may be inadequate for achieving the environmental sustainability and social equity goals of the movements. Unless groups working for change are aware of the need for deeper social restructuring, they may "end up legitimizing the very processes and interests they are seeking to change" (Clunies-Ross and Hildyard 1992: 8). Transformational perspectives and actions attempt to transcend incremental reformism and challenge the inequitable processes and structures at the core of the agrifood system itself. While incremental reformism can achieve some gains in the short run, it does not change the basic priorities and processes of the systems that created the problems in the first place. A key distinction can be made between which actions can be transformative (i.e., enhance present and future sustainability and food security for everyone) and which actions are accommodationist (i.e., allowing the abrogation of social responsibility for those less privileged).

---

2. These types of tensions persisted throughout the life of the project, culminating in the dissolution of the Alliance once the funding ended, although several organizations continued to collaborate on other projects.

Both alternative and oppositional efforts are needed, and it is possible to engage in both at the same time. The challenge lies in continuing to develop on-the-ground alternatives without losing sight of the big picture. Keeping the eye on the ball in this case means always recognizing that developing farmers' markets and CSAs or doing research on organic production methods are steps on the way of the longer path of creating an environmentally sound and socially just agrifood system. Groups working in isolation or on specific alternatives are unlikely to muster sufficient influence to drive significant change. For example, organizing around consumer issues has tended to focus on immediate, concrete disturbances and has seldom led to demands for larger changes or transformative shifts (Miller et al. 1995). Therefore, in addition to ad hoc organizing around specific issues, alternative agrifood movements need to connect food and agricultural problems with social and environmental issues in order to build a broad-based, vigorous, and powerful social movement. The problem for particular social movements is to "transcend particularities, and arrive at some conception of a universal alternative to that social system which is the source of their difficulties" (Harvey 2000).

Strengthening Alternative Agrifood Movements

The collective actions of alternative agrifood movements can be enhanced first by developing a broad-based vision for alternative agrifood systems that goes beyond traditional ideological and epistemological frameworks. An articulated vision of a sustainable and food secure society would help to engage and unite diverse constituents for an alternative agrifood movement. This is crucial because one of the fundamental requirements of a social movement is a problem statement and a way of expressing that problem—a clear discourse. A broader epistemological agenda, enabled by challenging current ideologies, would open new doors to how problems are conceptualized and which solutions are effected. Continuing to broaden constituencies and engage in democratic processes can provide the political power to move more quickly toward significant change in the agrifood system.

Alternatives to the Current Agrifood System

Taking time to actively reflect on the ideological constructions and social choices made and embedded within alternative agrifood concepts and strategies is a crucial step for building the alternative agrifood movement. We must do this if we are to resist ideologies, philosophies, epistemologies, and economic relations that set false limits on human possibilities or calcify "what is" as the model for "what should be." After all, social structures and systems do not exist outside or prior to the processes, flows, and relations that create, sustain, or undermine them (Harvey 1996).

Creating significant change requires understanding that the present organization of food and agriculture does not rest exclusively on any natural basis, but has developed as the product of power-laden human choices and their embodiment in social institutions. It is because the relationship between discourse and social structures is dialectical in this way that discourse assumes such importance in social change (Fairclough 2001). However, while institutions structure discourse, they do not *determine* it. New discourses, ideologies, and frameworks are being, can be, and need to be created in order to bring into being an environmentally sound and socially just agrifood system. This, in turn, requires an interrogation of the degree to which contemporary food and agriculture problems can be solved without examining and changing some existing social and economic relations. Systemic problems require systemic solutions, a basic tenet of the heuristic approach of political ecology, discussed in Chapter 5.

The agrifood system is socially organized, the outcome of everyday relationships among people and the institutions they create. All choices of priorities, action, and method reveal political and normative positions; whether they are expressed or latent, they are always present. Rather than accept present social and economic relations as preordained and inevitable, or normal and natural, alternative agrifood movements can envision new social and environmental arrangements that are compatible with meeting human needs and environmental sustainability. As Frances Moore Lappé (1990) reminds us, "We cannot move toward a future we cannot imagine, and we cannot imagine a future we don't believe is possible."

Articulating a Unified Vision

Alternatives to the current agrifood system should be based on a unified, concrete vision of who and what are to be sustained and secured. Current approaches tend to leave some fundamental questions unasked and therefore unanswered. The most fundamental is: Whom do we want to sustain and secure? How we choose to define these terms sets the parameters for what we include and what we exclude as problems as well as the strategies we employ to solve those problems. Claims that sustainable agriculture will benefit "society as a whole," for example, conceal very real class, gender, and ethnic divisions in American society. Additional fundamental questions include: What do we want to sustain—food production, groundwater levels, profits, existing gender relations? Who should benefit—family farmers, transnational food industries, local retailers, the hungry? What types of political-economic structures will facilitate the development of sustainable agriculture and food security—free market, planning, local decision making?

Sustainable agriculture has long been discussed in terms of the three "Es" of environment, economics, and equity as frameworks and goals. In further refining goals for an environmentally sound and socially just agrifood system, it may also be worth considering three "Es" to avoid. By this I mean destructive and familiar patterns of exploitation, extraction, and exclusion. In the category of exclusion I would place cultural patterns and forms such as gender, ethnicity, community, and locality. In exploitation, I include distributions of resource ownership and control, profits, wages, and entitlements. In the category of extraction I put "natural" resource issues such as the environment, soil, and water. The platform for sustainability and sustenance can—and should be—broad, but at the same time it needs to be as unambiguous as possible in providing defensible answers to these kinds of questions.

Expanding Participation

Developing a socially just and environmentally sound agrifood system must include representatives of all of the groups who participate in the agrifood system. This will require reaching beyond the class and color configurations of the traditional farm and environmental constituencies, both of which have been composed primarily of professional or self-employed European Americans. In contrast, most workers in the food manufacturing and service industries are nonunionized women, young people, and ethnic minorities (Friedmann 1995). Engaging consumers will also mean appealing to an incredibly diverse set of people.

Increasing the political efficacy of alternative agrifood movements will depend on building a broad constituency that includes all those who participate at various levels in the agrifood system. The only way to increase the numbers of alternative agrifood constituents is to appeal to a diversity of people, augmenting the emphasis on farmers and farming that currently exists in agrifood movements arenas (as discussed in Chapter 6). Although farmers and environmentalists are both important as supporters of efforts to develop alternative food systems, neither group is likely to provide the primary leadership of this movement (Buttel 1997). As Friedmann (1995: 24) points out, "Despite the high media profile of farmers, and their magical empowerment by molders of public opinion," new agents are more important to development of alternative agrifood policies. In today's agrifood system, numerical power lies not with farmers but with consumers and food workers in manufacturing and services. As stated earlier, alternative agrifood movements constitute a sort of discourse coalition. The power in this is that groups with similar—but not necessarily identical—interests can join together to achieve political objectives.

Building a cohesive movement requires subordinating particular interests at some levels, while effective coalitions and broad agendas are formed. The subordination of particular interests to collective interests requires a realignment of our frames of reference along both vertical and horizontal axes. A vertical analysis involves recognizing the links between and among conditions in food and agriculture, and understanding that environmental and social problems often have similar roots. A horizontal consciousness involves a redefinition of common interests, such as when consumers are interested not only in pesticide residues on food but also farmworker exposure to pesticides. The trick will be to form lasting alliances that build on *enough* of the similarities between sustainable agriculture and community food security proponents, along with workers, greens, feminists, and the poor, to provide a nourishing and dignified life on the earth's commons for everyone.

The Power of Food

Not every social problem generates a social movement; this potential lies with those issues that "strike a fundamental chord, that touch basic tensions in society" (Eyerman and Jamison 1991). Agrifood issues are clearly those that have the potential to catalyze broad social movements. Through its procurement, preparation, and consumption food structures some part of daily life for each of us. It is an "intimate" commodity in that it is something we take inside our bodies, which gives it special significance over commodities consumed outside the body (Winson 1993). A distinguishing feature of the food system from other systems of provision is a "preoccupation with the content of what we eat, whether it be nutritional, toxic and/or ecologically sound" (Fine and Leopold 1993: 149).[3] These characteristics of food mean that alternative agrifood movements have the potential to develop a broad-based social movement for change. The social movements of the 1960s recognized the power of food as a medium for broader social change. As Belasco (1989) writes, since people eat 365 days a year, three times a day, what could be more personal and political than food? As Friedmann (1995: 15) puts it, "Food and agriculture are enduring moments of social organization. . . . We all have to eat."

Concerns about the agrifood system span divisions of class and other social characteristics. There is an opportunity here to bring social justice and environmental justice together in these movements because of the universal need for food, which transcends identity, class, ethnicity, gender, and age. Wider participation will simultaneously require and lead to broader approaches to address and solve agrifood problems. Alternative agrifood movements have broken down the traditional production-consumption conceptual dualism in agrifood systems. This gives them the potential to bring together productive and distributive justice. This unity has implications for breaking down past divisions and creating new alliances among groups that have shared little common ground.

Alternative agrifood movements address issues at the very heart of human society, beginning with life itself. No other productive activity affects so many people in such a fundamental way as the one that produces and distributes our food. Nowhere are today's general trends of human and resource degradation clearer and their consequences sharper than in the agrifood system. These characteristics could also allow these movements to develop into even more encompassing movements for social and environmental justice. In the 1960s, for example, food activism was broad-based, and included fasts against the war, interracial dining at segregated restaurants, and consumer boycotts in support of agricultural workers, both domestic and international (Belasco 1989).

Participation in the movements need not mean becoming a full-time activist, researcher, or producer. People can participate effectively as consumers by changing their own perceptions and practices. The agency of

---

3. For example, three out of four Americans consider pesticide residues on food to be a serious health hazard (Food Marketing Institute 1993).

consumers acting within alternative agrifood efforts presents an opening for a significant restructuring and transformation of the agrifood system (Murdoch et al. 2000; Nygard and Storstad 1998; Whatmore and Thorne 1997). As Scott (1989: 5) points out, when everyday resistances are practiced widely, "they may have aggregate consequences out of all proportion to their banality when considered singly." Participation in everyday forms of resistance, like choosing foods grown without pesticides may seem small in comparison to the enormity of the problem, but they can have significant effects. Through food, the daily necessary practice of food provision and the objectives of social change can coincide. While green consumerism certainly has its limits, it can have political effects through the logic of the marketplace (Allen and Kovach 2000). Changes in diets can alter the structure of agrifood profits (Friedmann 1995) by, for example, increasing demand for foods grown without pesticides.

Furthermore, the critical thinking that is the basis of green consumer choices can come to include social justice issues such as the working conditions of agricultural laborers. In addition to consumption choices, other small acts can affect consciousness about the food system. Participating in a community garden or helping out at a food bank, for example, will unavoidably change some tiny part of how one sees the dynamics of the food system and perhaps see beyond it. Every action and every thought is significant. Every moment of every social relation presents the possibility of emancipatory or oppressive thought and action. Changes in consciousness inspired by concepts and practices of sustainable agriculture and community food security could lead to a consciousness that goes beyond the scope of the agrifood system, catalyzing a vigorous general movement for social and environmental justice.

There are additional institutional places in which the agrifood system could be more deeply democratized. For example, the Cornucopia Project (1981) included a recommendation that cities establish a "Department of Food" and that the United States develop a long-range national food plan. These ideas, which have been echoed in the Community Food Security movement, would provide excellent ways to include the priorities and perspectives of people in different social locations in the agrifood system. The effectiveness of people's participation would be enhanced if they knew even more about the agrifood system. For example, public programs to increase food literacy could be developed. A number of such efforts have been started in elementary schools, and food literacy programs and materials could also be developed for older students and for adults.

Another opportunity is public participation in research planning, proposal review, and outcome evaluation. While the USDA has long integrated advisory committees into its work, these have tended to represent commodity producers rather than a broad range of clientele, and are often used merely to rubber stamp decisions already made by administrators (Busch and Middendorf 1997). Public hearings have been used much the same way. However, the Land Grant Stakeholder Input Rule of 2000 requires colleges receiving federal formula funds for research to document how they gather and consider input from the public. Through this rule alternative agrifood groups have another avenue through which to affect the priorities and programs of traditional agrifood institutions.

One idea for widening the access to research resources is that of the "science shops" in Dutch universities. According to Busch and Middendorf (1997), these shops provide technical assistance to nonprofit groups who request it and provide access to scientific resources to those who otherwise might not have had access. A study by the Consortium for Sustainable Agriculture Research and Education recommended that more attention was needed to increase the participation of chronically underrepresented groups such as small-scale farmers, minority farmers, women, environmental groups, and consumer groups. These are precisely the kinds of groups have less ability to claim and provide resources in university programs, based on their limited resources and degree of cultural connections. This has implications for who are considered legitimate audiences of these programs and which kinds of research agendas are likely to be funded. Stevenson and others (1994) also suggest that research projects include representatives of organizations such as labor unions and citizens groups in the community of researchers. This would also be a way to foster deeper connections among practitioners, academics, and activists. These kinds of connections have always been a part of alternative agrifood movements, although there has at times been a disdain for the "ivory tower."

Nonetheless, those who are in professional functions as intellectuals can and do play important roles in social movements by helping to articulate concerns and placing them in broader frameworks, which can lend social actions deeper meaning or significance (Eyerman and Jamison 1991). Broader frameworks can be kept at the forefront by "organic" intellectuals, a concept developed by Antonio Gramsci, while he was imprisoned in Italy during the fascist era. Gramsci did not mean intellectual in an elitist sense, but rather counterposed "traditional" and "organic" intellectuals, the former being those who maintain the status quo and the latter those who challenge it. At the time of his writing, intellectuals were considered the enemies of the people. Gramsci's intent was to point out that the relevant distinction is not between intellectual and nonintellectual, but the purpose for which intellect is used. The organic intellectual is not necessarily a person; it can be a collective body that is able to articulate concepts and priorities of movements for social change. Organic intellectuals work toward forms of collective action that can bring into being a worldview that can help toward transformation of the structures that keep some people marginalized. The job of organic intellectuals is to study and frame social problems and put them in historical and political economic context. Perhaps these kinds of topics can be addressed through a virtual or face-to-face summit on sustainability and sustenance that would convene the different groups that constitute the American alternative agrifood movement.

Alternative agrifood institutions and movements are crucial to the future of humankind. As the twenty-first century begins, it is clear that we need fundamental change in the global food and agriculture system. Current conditions in food and agriculture—evidenced by deteriorating environmental, economic, and social circumstances—create the material need for and possibility of social transformation as embodied within alternative agrifood movements. Developing a food movement that works toward social and environmental justice requires developing a coherent vision that encompasses an understanding of the contradictions of the current system and includes all relevant constituencies in the movement. Alternative agrifood movements need to explicitly and critically address questions of who is included, who is left out, which problems are worthy of consideration, and which methods are appropriate in seeking solutions. Given the centrality of food for human existence, efforts to achieve sustainability and secure sustenance for everyone has to be comprehensive—simultaneously scientific and political, natural and social, based upon coherent theories and politically effective actions.

In the present moment, alternative agrifood movements are capable of becoming truly democratic and transformative, closing the gap between rhetoric and reality, principles and practice. Achieving agricultural sustainability and food security requires both the development of alternative practice and a political struggle over rights, justice, and equity. For Flacks (1995), the fate of democracy and the chances for social justice will depend on the capacity of social movements to take responsibility for the future. Alternative agrifood movements have this potential based on their potential to unite people from so many different locations and circumstances to create a socially just and environmentally sound agrifood system. We are all involved and we are all implicated. Harvey (1996: 106) asks, "So who and where are the agents of social change?" His answer is, "everyone, everywhere."

references

Adams, J. 1995. Individualism, efficiency, and domesticity: Ideological aspects of the exploitation of farm families and farm women. *Agriculture and Human Values* 12 (4): 2–17.

Adamson, M., and S. Borgos. 1984. *This mighty dream: Social protest movements in the United States*. Boston: Routledge & Kegan Paul.

Allen, P. 1999. Reweaving the food security safety net: Mediating entitlement and entrepreneurship. *Agriculture and Human Values* 16 (2): 117–29.

Allen, P., M. FitzSimmons, M. Goodman, and K. Warner. 2003. Shifting plates in the agrifood landscape: The tectonics of alternative agrifood initiatives in California. *Journal of Rural Studies* 19 (1): 61–75.

Allen, P., and M. Kovach. 2000. The capitalist composition of organic: The potential of markets in fulfilling the promise of organic agriculture. *Agriculture and Human Values* 17:221–32.

Allen, P., and D. Van Dusen. 1990. Raising fundamental issues. Sustainability in the Balance Issue Paper no. 1. Santa Cruz: Center for Agroecology and Sustainable Food Systems, University of California.

Altieri, M. A. 1987. *Agroecology: The scientific basis of alternative agriculture*. Boulder: Westview Press.

———. 1988. Beyond agroecology: Making sustainable agriculture part of a political agenda. *American Journal of Alternative Agriculture* 3:142–43.

Andrews, M. S., and K. L. Clancy. 1993. The political economy of the food stamp program in the United States. In *The political economy of food and nutrition policies*, ed. P. Pinstrup-Andersen. Baltimore: Johns Hopkins University Press.

Arcury, T. A., and Sara A. Quandt. 1998. Occupational and environmental health risks in farm labor. *Human Organization* 57 (3): 331–34.

Ashman, L., J. de la Vega, M. Dohan, A. Fisher, R. Hippler, and B. Romain. 1993. Seeds of change: Strategies for food security for the inner city. Los Angeles: Interfaith Hunger Coalition.

Atkinson, P. 1992. *Principles of political ecology*. London: Bellhaven Press.

Auburn, J., Director, USDA SARE Program. 2002. Telephone interview, 28 October.

Bacon, D. 1997. The UFW picks strawberries: Watsonville represents the largest, most intense organizing drive in America. *Nation* 264 (14): 18–22.

Baker, R. 1997. Where the sidewalks end. *Hope*, March/April, 16–23.

Barham, E. 1997. Social movements for sustainable agriculture in France: A Polanyian perspective. *Society & Natural Resources* 10:239–49.

Barlow, M., and T. Clarke. 1998. *MAI: The multilateral agreement on investment and the threat to American freedom*. New York: Stoddart.

Belasco, W. J. 1989. *Appetite for change*. New York: Pantheon Books.

Bell, D., and G. Valentine. 1997. *Consuming geographies*. London: Routledge.

Belsie, L. 1998. American farmers at the brink. *Christian Science Monitor*, 24 September, 1, 10.

Benton, T. 1989. Marxism and natural limits: An ecological critique and reconstruction. *New Left Review* 178 (November–December): 51–86.

Bernstein, J., H. Boushey, E. McNichol, and R. Zahradnik. 2002. Pulling apart: A state-by-state analysis of income trends. Washington, D.C.: Center on Budget and Policy Priorities and Economic Policy Institute.

Berry, W. 1988. *The unsettling of America*. San Francisco: Sierra Club Books.

Beus, C. E., and R. E. Dunlap. 1990. Conventional versus alternative agriculture: The paradigmatic roots of the debate. *Rural Sociology* 55 (4): 590–616.

Biehl, J., and M. Bookchin. 1998. *The politics of social ecology: Libertarian municipalism*. New York: Black Rose.

Bird, E. 1988. Why "modern" agriculture is environmentally unsustainable: Implications for the politics of the sustainable agriculture movement in the U.S. In *Global perspectives on agroecology and sustainable agricultural systems*, ed. P. Allen and D. Van Dusen, vol. 1. Santa Cruz: Agroecology Program, University of California.

Blanpied, Nancy A., ed. 1984. *Farm policy: The politics of soil, surpluses, and subsidies*. Washington, D.C.: Congressional Quarterly.

Boggs, C. 1986. *Social movements and political power: Emerging forms of radicalism in the west*. Philadelphia: Temple University Press.

Bourdieu, P., and L. J. D. Wacquant. 1992. *An invitation to reflexive sociology*. Chicago: University of Chicago Press.

Bowen, D. 1998. Crisis as opportunity. *California Certified Organic Farmers* 15 (2): 2–3.

Brown, K. H. 2002. Urban agriculture and community food security in the United States: Farming from the city center to the urban fringe. Ed. P. Mann. Report prepared by the Urban Agriculture Committee of the Community Food Security Coalition.

Browne, W. P. 1988. *Private interests, public policy, and American agriculture*. Lawrence: University Press of Kansas.

Browne, W. P., and A. J. Cigler, eds. 1990. *U.S. agricultural groups: Institutional profiles*. Westport, Conn.: Greenwood Press.

Browne, W. P., J. R. Skees, L. E. Swanson, P. B. Thompson, and L. J. Unnevehr. 1992. *Sacred cows and hot potatoes: Agrarian myths in agricultural policy*. Boulder: Westview Press.

Bryant, R. L. 1991. Putting politics first: The political ecology of sustainable development. *Global Ecology and Biogeography Letters* 1 (6): 164–66.

Buck, D., C. Getz, and J. Guthman. 1997. From farm to table: The organic vegetable commodity chain of Northern California. *Sociologia Ruralis* 37 (1): 4–20.

Busch, L., and W. B. Lacy. 1983. *Science, agriculture, and the politics of research*. Boulder: Westview Press.

Busch, L., and G. Middendorf. 1997. Democratic technology policy for a rapidly changing world. In *Visions of American Agriculture*, ed. W. Lockeretz. Ames: Iowa State University Press.

Buttel, F. H. 1980. Agriculture, environment, and social change: Some emergent issues. In *The rural sociology of advanced societies: Critical perspectives*, ed. F. H. Buttel and H. Newby. Montclair, N.J.: Allanheld, Osmun.

———. 1992. Environmentalization: Origins, processes, and implications for rural social change. *Rural Sociology* 57 (1): 1–27.

———. 1993a. The production of agricultural sustainability: Observations from the sociology of science and technology. In *Food for the future: Conditions and contradictions of sustainability*, ed. P. Allen. New York: John Wiley & Sons.

———. 1993b. The sociology of agricultural sustainability: Some observations on the future of sustainable agriculture. *Agriculture Ecosystems & Environment* 46 (1–4): 175–86.

———. 1997. The politics and policies of sustainable agriculture: Some concluding remarks. *Society & Natural Resources* 10:341–44.

———. 2000. The recombinant BGH controversy in the United States: Toward a new consumption politics of food? *Agriculture and Human Values* 17 (1): 5–20.

Buttel, F. H., and G. Gillespie, Jr. 1988. Agricultural research and development and the appropriation of progressive symbols: Some observations on the politics of ecological agriculture. Bulletin no. 151. Ithaca: Cornell University, Department of Rural Sociology.

Buttel, F. H., O. F. Larson, and G. W. Gillespie, Jr. 1990. *The sociology of agriculture*. Westport, Conn.: Greenwood Press.

California Farm Bureau Federation. 1998. California agriculture facts. Sacramento.

California Food and Agriculture Code. 1986. Sustainable agriculture research and education act of 1986. SB 872. *Statutes of 1986*. Sections 550–55.

———. 1994. Agricultural chemicals: Reduction pilot program. AB 3383. *Statutes of 1994*. Sections 591–600.

Campbell, D. 1997a. Community-controlled economic development as a strategic vision for the sustainable agriculture movement. *American Journal of Alternative Agriculture* 12 (1): 37–44.

———. 1997b. Welfare reform and community well-being: Public-private collaboration in California counties. Paper read at meeting of the Rural Sociological Society, 13–17 August, Toronto.

———. 2001. Conviction seeking efficacy: Sustainable agriculture and the politics of co-optation. *Agriculture and Human Values* 18:353–63.

Carter, H. O., and G. Goldman. 1996. *The measure of agriculture: Its impact on the state economy*. Oakland: Agricultural Issues Center, University of California.

Census Bureau. *See* U.S. Bureau of the Census.

Center on Budget and Policy Priorities. 1996. The depth of the food stamp cuts in the final welfare bill. Washington, D.C.: U.S. Government Printing Office.

Chisholm, A. H., and R. Tyers. 1982. Food security: An introduction and overview. In *Food security: Theory, policy, and perspectives from Asia and the Pacific Rim*, ed. A. H. Chisholm and R. Tyers. Lexington, Mass.: Lexington Books.

Chouliaraki, L., and N. Fairclough. 1999. *Discourse in late modernity: Rethinking critical discourse analysis.* Edinburgh: Edinburgh University Press.

Clancy, K. L. 1993. Sustainable agriculture and domestic hunger: Rethinking a link between production and consumption. In *Food for the future: Conditions and contradictions of sustainability,* ed. P. Allen. New York: John Wiley & Sons.

———. 1997. Reconnecting farmers and citizens in the food system. In *Visions of American agriculture,* ed. W. Lockeretz. Ames: Iowa State University Press.

Clunies-Ross, T., and N. Hildyard. 1992. *The politics of industrial agriculture.* London: Earthscan.

Cochrane, W. W. 1979. *The development of American agriculture: A historical analysis.* Minneapolis: University of Minnesota Press.

Cohen, J. L. 1985. Strategy or identity: New theoretical paradigms and contemporary social movements. *Social Research* 52 (4): 663–716.

Cohen, N. L., J. P. Cooley, R. B. Hall, and A. M. Stoddard. 1997. Community supported agriculture: A study of shareholders' dietary patterns and food practices. Paper presented at the International Conference on Agricultural Production and Nutrition, March, Tufts University, School of Nutrition Science and Policy, Boston.

Committee for Sustainable Agriculture. 1990. The Asilomar declaration for sustainable agriculture. Position paper presented to and endorsed at the Ecological Farming Conference, 9–11 January, Asilomar, California.

Community Food Security Coalition. 1994. *A community food security act: A proposal for new food system legislation as part of the 1985 farm bill.* By R. Gottlieb, A. Fisher, and M. Winne. Venice, Calif.: Community Food Security Coalition.

———. n.d. *The Healthy Farms, Food, and Communities Act: Policy initiatives for the 2002 Farm Bill and the first decade of the 21st century.* By A. Fisher, R. Gottlieb, T. Forster, and M. Winne. Venice, Calif.: Community Food Security Coalition.

Conway, G. R., and J. N. Pretty. 1991. *Unwelcome harvest: Agriculture and pollution.* London: Earthscan Publications.

Corbridge, S. 1998. Reading David Harvey: Entries, voices, loyalties. *Antipode* 30 (1): 43–55.

Cornucopia Project. 1981. *Empty breadbasket?* Emmaus, Pa.: Rodale Press.

Crews, T. E., C. L. Mohler, and A. G. Power. 1991. Energetics and ecosystem integrity: The defining principles of sustainable agriculture. *American Journal of Alternative Agriculture* 6:146–49.

Cromartie, J. B. 1999. Minority counties are geographically clustered. *Rural Conditions and Trends* 9 (2): 14–19.

Crosson, P. 1991. Sustainable agriculture in North America: Issues and challenges. *Canadian Journal of Agricultural Economics* 39:53–65.

Dahlberg, K. A. 1991. Sustainable agriculture—fad or harbinger? *BioScience* 41 (5): 337–40.

———. 1993. Regenerative food systems: Broadening the scope and agenda of sustainability. In *Food for the future: Conditions and contradictions of sustainability,* ed. P. Allen. New York: John Wiley & Sons.

———. 1994a. Food policy councils: The experience of five cities and one county. Unpublished manuscript.

———. 1994b. Alternative visions: Localizing food systems. *The Neighborhood Works*, February/March, 14.

Danbom, D. B. 1997. Past visions of American agriculture. In *Visions of American agriculture*, ed. W. Lockeretz. Ames: Iowa State University Press.

Darnovsky, M. 1992. Stories less told: Histories of U.S. environmentalism. *Socialist Review* 22 (4): 11–54.

Darnovsky, M., B. Epstein, and R. Flacks, eds. 1995. *Cultural politics and social movements*. Philadelphia: Temple University Press.

Davies, K., R. Wiles, and C. Campbell. n.d. Pay to spray. Web site of the Environmental Working Group. http://www.ewg.org/reports/PayToSpray/html.

de Janvry, A. 1980. Social differentiation in agriculture and the ideology of neopopulism. In *The rural sociology of advanced societies: Critical perspectives*, ed. F. H. Buttel and H. Newby. Montclair, N.J.: Allanheld, Osmun.

———. 1983. Why do governments do what they do? The case of food price policy. In *The role of markets in the world food economy*, ed. D. G. Johnson and G. E. Schuh. Boulder: Westview Press.

de Janvry, A., and E. P. LeVeen. 1986. Historical forces that have shaped world agriculture: A structural perspective. In *New directions for agriculture and agricultural research*, ed. K. A. Dahlberg. Totowa, N.J.: Rowman and Allanheld.

De Walt, B. R. 1985. Mexico's second Green Revolution: Food for feed. *Mexican Studies* 1 (1): 29–60.

DeLind, L. 1994. Organic Farming and Social Context: A Challenge for Us All. *American Journal of Alternative Agriculture* 9 (4): 146–47.

DeLind, L., and A. Ferguson. 1999. Is this a woman's movement? The relationship of gender to community-supported agriculture in Michigan. *Human Organization* 58:190–200.

DeLorme, C. D., Jr., D. R. Kamerschen, and D. C. Redman. 1992. The first U.S. Food Stamp Program: An example of rent seeking and avoiding. *American Journal of Economics and Sociology* 51 (4): 421–33.

Deshpande, S. 1991. To mould and harness: Capitalism, discipline, and discourse in the making of California agriculture. Ph.D. diss., Sociology Department, University of California, Santa Cruz.

Di Chiro, G. 1992. Defining environmental justice: Women's voices and grassroots politics. *Socialist Review* 22 (4): 93–130.

Dicks, M. R. 1992. What will be required to guarantee the sustainability of U.S. agriculture in the 21st century? *American Journal of Alternative Agriculture* 7 (4): 190–95.

Domhoff, G. W. 1991. State autonomy and new deal agricultural policy: Another empirical attack on a theoretical delusion. Paper presented at a meeting of the Pacific Sociological Association, 11–14 April, Irvine, Calif.

Douglass, G. K. 1984. The meanings of agricultural sustainability. In *Agricultural sustainability in a changing world order*, ed. G. K. Douglass. Boulder: Westview Press.

Dowler, E., and M. Caraher. 2003. Local food projects: The new philanthropy? *Political Quarterly* 74 (1): 57–65.

Dreze, J., and A. Sen. 1989. *Hunger and public action*. Oxford: Clarendon Press.

Duncan, C.A.M. 1996. *The centrality of agriculture*. Montreal: McGill-Queen's University Press.

Dunn, J. 1993. Freshman profiles. *Congressional Quarterly Weekly Report* 51 (3): 146.

Durning, A. B. 1990. Ending poverty. In *State of the World 1990*, ed. Worldwatch Institute. New York: W. W. Norton.

Dyer, J. 1999. In the eye of the stakeholder: Who sits at the agricultural research decision-making table? Web site of the Consortium for Sustainable Agriculture Research and Education. http://www.csare.org/pubs/dyer.htm. Last updated November 1999. on 17 December 2002.

Ebeling, W., ed. 1979. *The fruited plain: The story of American agriculture*. Berkeley and Los Angeles: University of California Press.

Economic Research Service. U.S. Department of Agriculture. 1985. *A profile of female farmers in America*. By J. Z. Kalbacher. Rural Development Research Report 45. Herndon, Va.: U.S. Department of Agriculture.

———. 1993. *Agricultural adaptation to urbanization: Farm types in the United States metropolitan areas*. By R. Heimlich and C. Barnard. Washington, D.C.: U.S. Department of Agriculture.

———. 1994. *Economic indicators of the farm sector: National financial summary 1992*. ECIFS 12-1. Herndon, Va.: Agriculture and Rural Economy Division.

———. 2001. *Agricultural income and finance outlook*. AIS-77. Washington, D.C.: U.S. Department of Agriculture. www.ers.usda.gov.

Edmondson, W. 2003. Food market structures: The U.S. food and fiber system. www.ers.usda.gov/Briefing/FoodMarketStructures/foodandfiber.htm. Last updated 1 April, 2003. Accessed on 19 August.

Edwards, C. A. 1990a. The importance of integration in sustainable agricultural systems. In *Sustainable agricultural systems*, ed. C. A. Edwards, R. Lal, P. Madden, R. H. Miller, and G. House. Ankeny, Iowa: Soil and Water Conservation Society.

———. 1990b. Preface. In *Sustainable agricultural systems*, ed. C. A. Edwards, R. Lal, P. Madden, R. H. Miller, and G. House. Ankeny, Iowa: Soil and Water Conservation Society.

Egan, T. 2000. Failing farmers learn to profit from federal aid. *New York Times*, December 24, p. 1.

Eisinger, P. 1996. Toward a national hunger count. *Social Service Review* 70 (2): 214–34.

Ellis, R. 1983. The way to a man's heart: Food in the violent home. In *The sociology of food and eating: Essays on the sociological significance of food*, ed. A. Murcott. Aldershot, Hants, England: Gower.

Elswick, L., and T. Forster. 2002. Community-based food security comes to the world stage in a country near you! *Community Food Security News* (Venice, Calif., newsletter of Community Food Security Coalition), Spring: 9.

Engberg, L. 1996. Livelihood and food security: Issues for women and families. NGO Comments on WFS Policy and Plan of Action. Ottowa, Ontario: United Nations Association in Canada. http://www.unac.org.gnfs/ngolila.htm. Accessed on 5 September 1998.

Environmental Working Group. 2000. New EPA study elevates cancer rating for top U.S. weed killer. Washington, D.C.: Environmental Working Group.

Environmental Working Group California. 1999. Tap water in 38 Central California cites tainted with banned pesticide: Some bottle-fed infants may exceed "safe" dose before age 1. San Francisco: Environmental Working Group California.

Epstein, B. 1990. Rethinking social movement theory. *Socialist Review* 20 (1): 35–65.

Esbjornson, C. D. 1992. Once and future farming: Some meditations on the historical and cultural roots of sustainable agriculture in the United States. *Agriculture and Human Values* 9 (3): 20–30.

Eyerman, R., and A. Jamison. 1991. *Social movements: A cognitive approach.* University Park: The Pennsylvania State University Press.

Faeth, P., R. Repetto, K. Kroll, Q. Dai, and G. Helmers. 1991. *Paying the farm bill: U.S. agricultural policy and the transition to sustainable agriculture.* Washington, D.C.: World Resources Institute.

Fairclough, N. 1994. *Discourse and social change.* Cambridge: Polity Press.

———. 2001. *Language and power.* 2d ed. Essex: Pearson Educational Limited.

Featherstone, L., and United Students Against Sweatshops. 2002. *Students against sweatshops.* New York: Verso.

Feenstra, G. W. 1997. Local food systems and sustainable communities. *American Journal of Alternative Agriculture* 12 (1): 28–36.

Feenstra, G., and D. Campbell. 1998. Community food systems in California. University of California Sustainable Agriculture Research and Education Program Publication 21574. Oakland: Division of Agriculture and Natural Resources.

Feldman, S., and R. Welsh. 1995. Feminist knowledge claims, local knowledge, and gender divisions of agricultural labor: Constructing a successor science. *Rural Sociology* 60 (1): 23–42.

Festing, H. 1997. Community supported agriculture and vegetable box schemes. Paper presented at the International Conference on Agricultural Production and Nutrition, March, Tufts University, School of Nutrition Science and Policy, Boston.

Fine, B., and E. Leopold. 1993. *The world of consumption.* London: Routledge.

Fink, D. 1992. *Agrarian women.* Chapel Hill: University of North Carolina Press.

Fisher, A., Executive Director, Community Food Security Coalition. 1997. What is community food security? *Urban Ecologist* 2:3–4.

———. 1998. Email to coalition members, 24 November.

———. 2002a. Telephonic interview by author. 31 October.

———. 2002b. Community food security: A promising alternative to the global food system. *Community Food Security News* (Venice, Calif., newsletter of Community Food Security Coalition), Spring: 5.

Fisher, A., and R. Gottlieb. 1995. Community food security: Policies for a more sustainable food system in the context of the 1995 farm bill and beyond. Working Paper no. 11. The Lewis Center for Regional Policy Studies. School of Public Policy and Social Research. University of California, Los Angeles.

Fisk, J. W., O. B. Hesterman, and T. L. Thorburn. 1998. Integrated farming systems in the USA. In *Facilitating sustainable agriculture*, ed. N. G. Röling and M.A.E. Wagemakers. Oxford: Oxford University Press.

Fitchen, J. M. 1997. Hunger, malnutrition, and poverty in the contemporary United States. In *Food and culture: A Reader*, ed. C. Counihan and P. V. Esterik. New York: Routledge.

FitzSimmons, M. 1990. The social and environmental relations of U.S. agricultural regions. In *Technological change and the rural environment*, ed. P. Lowe, T. Marsden, and S. Whatmore. London: David Fulton Publishers.

Flacks, R. 1995. Think globally, act politically: Some notes toward new movement strategy. In *Cultural politics and social movements*, ed. M. Darnovsky, B. Epstein, and R. Flacks. Philadelphia: Temple University Press.

Food and Agriculture Organization. 1996. *FAO trade yearbook*. FAO Statistics Series, 132. Rome: Food and Agriculture Organization.

Food First Information and Action Network. Institute for Food and Development Policy. 1997. The 12 misconceptions about the right to food. http://www.foodfirst.org/miscon.htm. Accessed on 5 September 1998.

Food Marketing Institute. 1993. Consumer attitudes and the supermarket. http://www.monitor.net/~cap/org_consumers.html. Last updated 1993. Accessed on 7 January 2002.

Food Research and Action Center. 1995. A survey of childhood hunger in the United States. Washington, D.C.: Food Research and Action Center, July.

Food security internationally. 1997. *Urban Ecologist*, no. 2: 4.

Foster, J. B. 1997. Ecocommunism: Ecological sustainability in Marx's vision of future society. Paper presented at the conference "What Is Ecological Socialism?" 9–11 May, Santa Cruz, Calif.

Foster, J. B., and F. Magdoff. 1998. Liebig, Marx, and the depletion of soil fertility: Relevance for today's agriculture. *Monthly Review* 50 (3): 32–45.

Fox, G. 1990. The economics of the sustainable agriculture movement. *Canadian Journal of Agricultural Economics* 38:727–39.

Francis, C. A. 1988. Research and extension for sustainable agriculture. *American Journal of Alternative Agriculture* 3:123–26.

Francis, C., G. Lieblein, S. Gliessman, T. A. Breland, N. Creamer, R. Harwood, L. Salomonsson et al. 2003. Agroecology: The ecology of food systems. *Journal of Sustainable Agriculture* 22 (3): 99–118.

Frank, A. G., and M. Fuentes. 1990. Civil democracy: Social movements in recent world history. In *Transforming the revolution: Social movements and the world-system*, ed. S. Amin, G. Arrigihi, A. G. Frank, and I. Wallerstein. New York: Monthly Review Press.

Frenkel, S. 1994. Environmental determinism and bioregionalism. *Professional Geographer* 46 (3): 289–95.

Freudenberger, C. D. 1986. Value and ethical dimensions of alternative agricultural approaches: In quest of a regenerative and just agriculture. In *New directions for agriculture and agricultural research*, ed. K. A. Dahlberg. Totowa, N.J.: Rowman and Allanheld.

Friedland, W. H. 1991. Women and agriculture in the United States: A state of the art assessment. In *Towards a new political economy of agriculture*, ed. W. H. Friedland, L. Busch, F. H. Buttel, and A. P. Rudy. Boulder: Westview Press.

Friedland, W. H., F. H. Buttel, and A. P. Rudy. 1991. Introduction: Shaping the new political economy of advanced capitalist agriculture. In *Towards a new political economy of agriculture*, ed. W. H. Friedland, L. Busch, F. H. Buttel, and A. P. Rudy. Boulder: Westview Press.

Friedmann, H. 1993. After Midas' feast: Alternative food regimes for the future. In *Food for the future: Conditions and contradictions of sustainability*, ed. P. Allen. New York: John Wiley & Sons.

———. 1994. Distance and durability: Shaky foundations of the world food economy. In *The global restructuring of agro-food systems*, ed. P. McMichael. Ithaca: Cornell University Press.

———. 1995. Food politics: New dangers, new possibilities. In *Food and agrarian orders in the world-economy*, ed. P. McMichael. Westport, Conn.: Praeger.

Friedmann, H., and P. McMichael. 1989. Agriculture and the state system: The rise and decline of national agricultures, 1870 to the present. *Sociologia Ruralis* 29 (2): 93–117.

Funders Agricultural Working Group. 2001. Roots of change: Agriculture, ecology, and health in California. San Francisco: Funders Agricultural Working Group.

Fuster, E., Director of Urban Agriculture, Havana, Cuba. 1998. Seminar on Cuba's urban agriculture, University of California, Santa Cruz, 5 October.

Gal, S. 1992. Language, gender, and power: An anthropological view. In *Locating power: Proceedings of the Second Berkeley Women and Language Conference, April 4 and 5, 1992*, ed. M. B. Kira Hall, Birch Moonwomon. Berkeley: Berkeley Women and Language Group, University of California, Berkeley.

Gans, H. J. 1995. *The war against the poor*. New York: Basic Books.

GAO. *See* U.S. General Accounting Office.

Gardner, B. 1983. Discussion [of de Janvry, Alain. 1983. Why do governments do what they do? The case of food price policy]. In *The role of markets in the world food economy*, ed. D. G. Johnson and G. E. Schuh. Boulder: Westview Press.

Geithman, F. E., and B. Marion. 1993. Testing for market power in supermarket prices: A review of the Kaufman-Handy/ERS study. In *Competitive strategy analysis in the food system*, ed. R. W. Cotterill. Boulder: Westview Press.

George, S. 1985. Rejoinder: On the need for a broader approach. *Food Policy* 10 (1): 75–79.

Gershuny, G., and T. Forster. 1992. Does organic mean socially responsible? A conversation. *Organic Farmer* 3 (1): 7–11.

Gilg, A., and M. Battershill. 1998. Quality farm food in Europe: A possible alternative to the industrialised food market and to current agri-environmental policies: Lessons from France. *Food Policy* 23 (1): 25–40.

Gips, T. 1988. What is sustainable agriculture? In *Global perspectives on agroecology and sustainable agricultural systems*, ed. P. Allen and D. Van Dusen, vol. 1. Santa Cruz: University of California.

Goldberg, R. A. 1991. *Grassroots resistance: Social movements in twentieth-century America*. Belmont, Calif.: Wadsworth Publishing.

Goldring, L. 1996. Gendered memory: Constructions of rurality among Mexican transnational migrants. In *Creating the countryside: The politics of rural and environmental discourse*, ed. E. M. DuPuis and P. Vandergeest. Philadelphia: Temple University Press.

Goodman, D. 1999. Agro-food studies in the "Age of Ecology": Nature, corporeality, bio-politics. *Sociologia Ruralis* 39 (1): 17–38.

Goodman, R. 1997. Ensuring the scientific foundations for America's future. In *Visions of American Agriculture*, ed. W. Lockeretz. Ames: Iowa State University Press.

Gorham, L. 1992. The growing problem of low earnings in rural areas. In *Rural poverty in America*, ed. C. M. Duncan. New York: Auburn House.

Gottlieb, R. 2001. *Environmentalism unbound: Exploring new pathways for change.* Cambridge, Mass.: MIT Press.

———. 2003. Community food security: What is it about? What brings us together? *Community Food Security News* (Venice, Calif., newsletter of the Community Food Security Coalition), Fall 2002–Winter 2003: 6–7.

Gottlieb, R., and A. Fisher. 1996a. Community food security and environmental justice: Searching for a common discourse. *Agriculture and Human Values* 3 (3): 23–32.

———. 1996b. "First feed the face": Environmental justice and community food security. *Antipode* 28 (2): 193–203.

———. 1998. Community food security and environmental justice: Converging paths toward social justice and sustainable communities. *Community Food Security News* (Venice, Calif., newsletter of the Community Food Security Coalition), Summer: 4–5.

Gottlieb, R., and H. Joseph. 1997. *Building toward the millennium: Understanding the past and envisioning the future of the Community Food Security Coalition.* Discussion paper distributed at first National Community Food Security Conference, 25–26 October, Los Angeles.

Grey, M. 2000. The industrial food stream and its alternatives in the United States: An introduction. *Human Organization* 59 (2): 143–50.

Grieshop, J. I., K. Peck, and A. Raj. 1996. Marketing sustainable agriculture: A promoter's toolbox. Publication 3367. Oakland: University of California, Division of Agriculture and Natural Resources.

Guither, H. D. 1980. *The food lobbyists.* Lexington, Mass.: Lexington Books.

Guthman, J. 1998. Regulating organic: Food safety, environmental risk, and the political economy of agriculture. Paper presented at the annual meeting of the Rural Sociological Society, 6–9 August, Portland, Ore.

Hadwiger, D. F., and W. P. Browne, eds. 1987. *Public policy and agricultural technology: Adversity despite achievement.* Policy Studies Organization Series. Ed. S. S. Nagel. London: Macmillan Press in association with the Policy Studies Organization.

Haila, Y., and R. Levins. 1992. *Humanity and nature: Ecology, science, and society.* London: Pluto Press.

Hajer, M. A. 1995. *The politics of environmental discourse.* Oxford: Clarendon Press.

Hall, S. 1982. The rediscovery of "ideology": Return of the repressed in media studies. In *Culture, society and the media*, ed. M. Gurevitch, T. Bennett, J. Curran, and J. Woollacott. London: Methuen.

Hallberg, M. C. 2001. *Economic trends in U.S. agriculture since World War II*. Ames: Iowa State University Press.

Hamel, K. 1995. Harvesting handouts l: The federal farm price support scandal. Washington, D.C.: Public Voice for Food and Health Policy.

Hamlin, C. 1991. Green meanings: What might "sustainable agriculture" sustain? *Science as Culture* 2 (13): 507–37.

Hansen, V. 1999. Farmers harvest a bumper crop of subsidies. *Wall Street Journal*, August 10, A24.

Havens, A. E. 1986. Capitalist development in the United States: State, accumulation, and agricultural production systems. In *Studies in the transformation of U.S. agriculture*, ed. A. E. Havens. Boulder: Westview Press.

Havens, A. E., and H. Newby. 1986. Agriculture and the state: An analytical approach. In *Studies in the transformation of U.S. agriculture*, ed. A. E. Havens. Boulder: Westview Press.

Harvey, D. 1996. *Justice, nature and the geography of difference*. Cambridge: Blackwell.

———. 2000. *Spaces of hope*. Edinburgh: Edinburgh University Press.

Hayward, T. 1992. Ecology and human emancipation. *Radical Philosophy* 62:3–12.

———. 1994. The meaning of political ecology. *Radical Philosophy* 66:11–20.

Hassanein, N. 1999. *Changing the way America farms: Knowledge and community in the sustainable agriculture movement*. Lincoln: University of Nebraska Press.

Healy, R. G., and J. L. Short. 1981. *The market for rural land: Trends, issues, policies*. Washington, D.C.: The Conservation Foundation.

Henderson, E. 1998. Rebuilding local food systems from the grassroots up. *Monthly Review* 50 (3): 112–24.

Hendrickson, M., President, Community Food Security Coalition board of directors. 2002. Interview, 27 September.

Hendrickson, M., W. D. Heffernan, P. H. Howard, and J. B. Heffernan. 2001. Consolidation in food retailing and dairy: Implications for farmers and consumers in a global food system. Columbia: University of Missouri.

Hera, C. 1995. Atoms for sustainable agriculture: Enriching the farmer's field. *IAEA Bulletin* 2:36–41.

Herrin, M., and J. D. Gussow. 1989. Designing a sustainable regional diet. *Journal of Nutrition Education* 21 (6): 270–75.

Herring, S., D. A. Johnson, and T. DiBenedetto. 1995. "This discussion is going too far!" Male resistance to female participation on the Internet. In *Gender articulated: Language and the socially constructed self*, ed. K. Hall and M. Bucholtz. New York: Routledge.

Hesterman, O. B., and T. L. Thorburn. 1994. A comprehensive approach to sustainable agriculture: W. K. Kellogg's integrated farming systems initiative. *Journal of Production Agriculture* 7 (1): 132–43.

Hinrichs, C. 2000. Embeddedness and local food systems: Notes on two types of direct agricultural markets. *Journal of Rural Studies* 16 (3): 295–303.

————.2003. The practice and politics of food system localization. *Journal of Rural Studies* 19 (1): 33–45.

Hinrichs, C., and K. Kremer. 1998. The challenges of class for community supported agriculture: Emerging insights in Central Iowa. Paper presented at the joint annual meeting of the Association for the Study of Food and Society and the Agriculture, Food, and Human Values Society, 4–6 June, San Francisco.

————. 2002. Social inclusion in a Midwest local food system project. *Journal of Poverty* 6 (1): 65–90.

Hosansky, D. 1995. Reconciliation: Clashing needs: Budget cuts vs. money for food stamps. *Congressional Quarterly Weekly Report* 53 (43): 3363–65.

House. *See* U.S. Congress. House.

Howard, A. 1943. *An agricultural testament.* London: Oxford University Press.

Hunter, A. 1995. Rethinking revolution in light of the new social movements. In *Cultural politics and social movements*, ed. M. Darnovsky, B. Epstein, and R. Flacks. Philadelphia: Temple University Press.

Hynes, H. P. 1996. *A patch of Eden: America's inner-city gardeners.* White River Junction, Vt.: Chelsea Green.

Ilbery, B., and M. Kneafsey. 1999. Niche markets and regional specialty food products in Europe: Towards a research agenda. *Environment and Planning A* 31 (12): 2207–22.

Inouye, J., and K. D. Warner. 2001. Plowing ahead: Working social concerns into the sustainable agriculture movement. A CalsAWG White Paper. Santa Cruz: California Sustainable Agriculture Working Group.

Integrated Food and Farming Systems Network. 1998. *Network News* 1 (1).

International Labour Office. 1988. *Wages in agriculture.* Geneva: International Labour Office.

International Movement for Ecological Agriculture. 1991. Declaration: From global crisis towards ecological agriculture. *Lokayan Bulletin* 9 (1): 78–87.

International Planning Committee for the World Food Summit. 2002. Profit for few or food for all. Rome: NGO/CSO Forum for Food Sovereignty. Available on-line at http://itdg.org/html/advocacy/wfs.htm.

Jackson, C. 1994. Gender analysis and environmentalisms. In *Social theory and the global environment*, ed. M. Redcliff and T. Benton. London: Routledge.

Jackson, W. 1980. *New roots for agriculture.* San Francisco: Friends of the Earth.

————. 1990. Agriculture with nature as analogy. In *Sustainable agriculture in temperate zones*, ed. C. A. Francis, C. B. Flora, and L. D. King. New York: John Wiley & Sons.

Jelinek, L. J. 1982. *Harvest empire: A history of California agriculture.* 2d ed. San Francisco: Boyd & Fraser.

Kansas City Star. 1991. Inside the USDA, it's a white male bastion. 8–14 December.

Kantor, L. S. 2001. Community food security programs improve food access. *Food Review* 24 (1): 20–26.

Kautsky, K. 1988. *The agrarian question.* 2 vols. London: Zwan Publications.

Kearney, M., and C. Nagengast. 1989. Anthropological perspectives on transnational communities in rural California. Working Group on Farm Labor and

Rural Poverty Working Paper no. 3. Davis: California Institute for Rural Studies.

Keister, L. A. 2000. *Wealth in America: Trends in wealth inequality*. Cambridge: Cambridge University Press.

Kellogg. *See* W. K. Kellogg Foundation.

Kemmis, D. 1990. *Community and the politics of place*. Norman: University of Oklahoma Press.

Kirkby, J., P. O'Keefe, and L. Timberlake. 1995. *The Earthscan reader in sustainable development*. London: Earthscan Publications.

Kirshenmann, F. 2002. Why American agriculture is not sustainable. *Renewable Resources Journal* 20 (3): 6–11.

Kloppenburg, J., Jr. 1991. Alternative agriculture and the new biotechnologies. *Science as Culture* 2 (13): 482–506.

Kloppenburg, J., Jr., J. Hendrickson, and G. W. Stevenson. 1996. Coming into the foodshed. *Agriculture and Human Values* 13 (3): 33–42.

Kloppenburg, J., S. Lezberg, K. De Master, G. Stevenson, and J. Hendrickson. 2000. Tasting food, tasting sustainability: Defining the attributes of an alternative food system with competent, ordinary people. *Human Organization* 59 (2): 177–86.

Knutson, R. D., J. B. Penn, and B. L. Flinchbaugh. 1998. *Agricultural and food policy*. Upper Saddle River, N.J.: Prentice Hall.

Koc, M. 1994. Globalization as a discourse. In *From Columbus to ConAgra: The globalization of agriculture and food*, ed. A. Bonanno, L. Busch, W. Friedland, L. Gouveia, and E. Mingione. Lawrence: University Press of Kansas.

Kolodinsky, J., and L. Pelch. 1997. Factors influencing the decisions to join a community supported agriculture farm. *Journal of Sustainable Agriculture* 10 (2/3): 129–41.

Kotz, N. 1969. *Let them eat promises: The politics of hunger in America*. Garden City, N.Y.: Anchor Books.

Krebs, A. V. 1991. *The corporate reapers*. Washington, D.C.: Essential Books.

Kuminoff, N., D. Sumner, and G. Goldman. 2000. *The measure of California agriculture, 2000*. Davis: University of California Agricultural Issues Center.

Lacy, W. 2000. Empowering communities through public work, science, and local food systems: Revisiting democracy and globalization. *Rural Sociology* 65 (1): 3–26.

Lappé, F. M. 1990. Food, farming, and democracy. In *Our sustainable table*, ed. R. Clark. San Francisco: North Point Press.

Lappé, F. M., and R. Schurman. 1988. The missing piece in the population puzzle. Food First Development Report no. 4. San Francisco: Institute for Food and Development Policy.

Lawrence, K. 1997. A legacy of csa and a vision for world peace. *Community Food Security News* (Venice, Calif., newsletter of the Community Food Security Coalition), Winter: 5.

Lawson, J. E. 1997. This land shall be forever stewarded: A story of a community's effort to preserve the farm through sharing property equity. Davis, Calif.: Community Alliance with Family Farmers.

Lehman, H., E. A. Clark, and S. F. Wise. 1993. Clarifying the definition of sustainable agriculture. *Journal of Agricultural and Environmental Ethics* 6 (2): 127–43.

Leon, W., and C. Smith DeWaal. 2002. *Is our food safe?* New York: Three Rivers Press.

Lewis, M. W., ed. 1992. Green delusions: An environmentalist critique of radical environmentalism. Durham: Duke University Press.

Lezberg, S., and J. Kloppenburg, Jr. 1996. That we all might eat: Regionally-reliant food systems for the 21st century. *Development* 4:28–33.

Liebhardt, B. 1997. SAREP refines mission, goals. *Sustainable Agriculture* 9 (2): 1.

Lipietz, A. 1995. *Green hopes: The future of political ecology.* Cambridge: Polity Press.

Lipman-Blumen, J. 1986. Exquisite decisions in a global village. In *New directions for agriculture and agricultural research*, ed. K. A. Dahlberg. Totowa, N.J.: Rowman and Allanheld.

Lipset, S. M. 1968. *Agrarian socialism.* Garden City, N.Y.: Anchor Books.

Lipsky, M., and M. A. Thibodeau. 1990. Domestic food policy in the United States. *Journal of Health Politics* 15 (2): 319–39.

Lobao, L. M., and P. Thomas. 1992. Political beliefs in an era of economic decline: Farmers' attitudes toward state economic intervention, trade, and food security. *Rural Sociology* 57 (4): 453–75.

Lockeretz, W. 1988. Open questions in sustainable agriculture. *American Journal of Alternative Agriculture* 3:174–81.

———. 1989. Comparative local economic benefits of conventional and alternative cropping systems. *American Journal of Alternative Agriculture* 4 (7): 75–83.

———. 1991. Multidisciplinary research and sustainable agriculture. *Biological Agriculture & Horticulture* 8 (2): 101–22.

Lockeretz, W., and M. D. Anderson. 1993. *Agricultural research alternatives.* Lincoln: University of Nebraska Press.

Lukes, S. 1975. *Power: A radical view.* London: Macmillan.

Luttrell, C. B. 1989. *The high cost of farm welfare.* Washington, D.C.: Cato Institute.

Lyson, T. A., G. W. Gillespie, Jr., and D. Hilchey. 1995. Farmers' markets and the local community: Bridging the formal and informal economy. *American Journal of Alternative Agriculture* 10 (3): 108–13.

MacCannell, D. 1988. Industrial agriculture and rural community degradation. In *Agriculture and community change in the U.S.*, ed. L. E. Swanson. Boulder: Westview Press.

Madden, J. P. 1998. The early years of the LISA, SARE, and ACE programs. Web site of Western Regional Sustainable Agricultural Research and Education. http://wsare.usus.edu/history/docinfo.htm. Accessed on 30 April 2002.

Magdoff, F., J. B. Foster, and F. H. Buttel. 1998. Introduction. *Monthly Review* 50 (3): 1–13.

Maggard, S. W., and R. B. Thompkins. 1998. Gender, race, and class in enterprise communities: Inherited social capital and persistent inequality. Paper presented at the annual meeting of the Rural Sociological Society, 8 August, Portland, Ore.

Makinson, L., and J. Goldstein, eds. 1994. *The cash constituents of Congress.* Washington, D.C.: Congressional Quarterly.

Mann, S. A., and J. A. Dickinson. 1980. State and agriculture in two eras of American capitalism. In *The rural sociology of the advanced societies: Critical perspectives*, ed. F. H. Buttel and H. Newby. Montclair, N.J.: Allanheld, Osmun.

Marsden, T. 2000. Food Matters and the Matter of Food: Towards a New Food Governance? *Sociologia Ruralis* 40:20–29.

Marsden, T., and A. Arce. 1995. Constructing quality: Emerging food networks in the rural transition. *Environment and Planning A* 27 (8): 1261–79.

Marsden, T., and J. Little. 1990. Introduction. In *Political, social, and economic perspectives on the international food system*, ed. T. Marsden and J. Little. Aldershot, England: Avebury.

Martin, P. L., S. Vaupel, and D. L. Egan. 1988. *Unfilled promise: Collective bargaining in California agriculture*. Boulder: Westview Press.

Martinez-Alier, J. 1995. Political ecology, distributional conflicts, and economic incommensurability. *New Left Review* (211): 70–88.

Marx, K. 1976. *Capital*. Volume 1. New York: Vintage.

McAdam, D., J. D. McCarthy, and M. N. Zald. 1988. Social movements. In *The handbook of sociology*, ed. N. J. Smelser. Beverly Hills, Calif.: Sage Publications.

McCalla, A. 1978. The politics of the U.S. agricultural research establishment: A short analysis. *Policy Studies Journal* 4 (4).

McCalla, A. F., and E. Learn. 1985. Public policies for food, agriculture, and resources: Retrospect and prospect. In *The dilemmas of choice*, ed. K. A. Price. Washington, D.C.: Resources for the Future.

McCann, T. 1998. Food fight: National organic standards. *MFA Digest* 12 (2): 1–2.

McConnell, G. 1959. *The decline of agrarian democracy*. Berkeley and Los Angeles: University of California Press.

McGinnis, J. M., and W. H. Foege. 1993. Actual causes of death in the United States. *Journal of the American Medical Association* 270 (18): 2207–12.

McMichael, P. 2000. The power of food. *Agriculture and Human Values* 17:21–33.

McWilliams, C. 1939. *Factories in the field*. Boston: Little, Brown.

Meares, A. C. 1997. Making the transition from conventional to sustainable agriculture: Gender, social movement participation, and quality of life on the family farm. *Rural Sociology* 62 (1): 21–47.

Merchant, C. 1992. *Radical ecology: The search for a livable world*. London: Routledge, Chapman & Hall.

Merrigan, K. A. 1993. National policy options and strategies to encourage sustainable agriculture: Lessons from the 1990 farm bill. *American Journal of Alternative Agriculture* 8 (4): 158–61.

———. 1997. Government pathways to true food security. In *Visions of American agriculture*, ed. W. Lockeretz. Ames: Iowa State University Press.

Metcalf, R. L., and W. H. Luckmann. 1982. *Introduction to insect pest management*. 2d ed. New York: John Wiley & Sons.

Milio, N. 1991. Food rich and health poor. *Food Policy* 16 (4): 311–18.

Miller, L. C., and M. Neth. 1988. Farm women in the political arena. In *Women and farming: Changing roles, changing structures*, ed. W. G. Haney and J. B. Knowles. Boulder: Westview Press.

Miller, S. M., M. Rein, and P. Levitt. 1995. Community action in the United States. In *Community empowerment*, ed. G. Craig and M. Mayo. London: Zed Books.

Mills, F. B. 1996. The ideology of welfare reform: Deconstructing stigma. *Social Work* 41 (4): 391–96.

Mintz, S. 1995. Food and its relationship to concepts of power. In *Food and agrarian orders in the world-economy*, ed. P. McMichael. Westport, Conn.: Praeger.

Mooney, P. H., and T. J. Majka. 1995. *Farmers' and farm workers' movements: Social protest in American agriculture*. New York: Twayne Publishers.

Moore, D. S. 1996. Marxism, culture, and political ecology. In *Liberation ecologies: Environment, development, and social movements*, ed. R. Peet and M. Watts. London: Routledge.

Murdoch, J., T. Marsden, and J. Banks. 2000. Quality, nature, embeddedness: Some theoretical considerations in the context of the food sector. *Economic Geography* 76 (2): 107–18.

Murdoch, J., and M. Miele. 1999. "Back to nature": Changing "worlds of production" in the food sector. *Sociologia Ruralis* 39:465–83.

Nachman-Hunt, N. 2002. USDA certified organic: Field of dreams? *Lohas Journal* 3 (3).

Naples, N. 1994. Contradictions in agrarian ideology: Restructuring gender, race, ethnicity and class. *Rural Sociology* 59 (1): 110–34.

National Center for Chronic Disease Prevention and Health Promotion. 2000. Prevalence of obesity among US adults, by characteristic. http://www. cdc.gov/nccdphp/dnpa/obesity/prevtable91-99char.htm. Accessed on 13 March 2001.

National Commission on the Environment. 1993. Choosing a sustainable future: The report of the National Commission on the Environment. Washington, D.C.: Island Press.

National Institute of Occupational Safety and Health. 2002. *Worker health chartbook, 2000*. Cincinnati, Ohio: National Institute of Occupational Safety and Health.

National Research Council. Board on Agriculture. National Academy of Sciences. 1989. *Alternative agriculture*. Washington, D.C.: National Academy Press.
———. 1995. *Colleges of agriculture and the land-grant universities: A profile*. Washington, D.C.: National Academy Press.

National Science Foundation. 1989. *Profiles—agricultural science: Human resources and funding*. NSF 89-319. Washington, D.C.: National Science Foundation.

National Sustainable Agriculture Coordinating Council. n.d. The campaign for sustainable agriculture: Working toward a new direction in federal farm and food policy. Goshen, N.Y.: National Sustainable Agriculture Coordinating Council.

Nestle, M. 1993. Food lobbies, the food pyramid, and United States nutrition policy. *International Journal of Health Services* 23 (3): 483–96.

Nestle, M., and S. Guttmacher. 1992. Hunger in the United States: Rationale, methods, and policy implications of state hunger surveys. *Journal of Nutrition Education* 24 (1): 18S–22S.

Neuhauser, L., D. Disbrow, and S. Margen. 1995. *Hunger and food insecurity in California*. Berkeley: California Policy Seminar, University of California.

Norgaard, R. B., ed. 1994. Development betrayed: The end of progress and a coevolutionary revisioning of the future. London: Routledge.

NRC. *See* National Research Council.

Nygard, B., and O. Storstad. 1998. De-globalization of food markets? Consumer perceptions of safe food: The case of Norway. *Sociologia Ruralis* 38 (1): 35–53.

O'Connor, J. 1993. Is sustainable capitalism possible? In *Food for the future: Conditions and contradictions of sustainability*, ed. P. Allen. New York: John Wiley & Sons.

———. 1998. *Natural causes: Essays in ecological Marxism*. New York: Guilford.

O'Hare, W. P. 1988. *The rise in poverty in rural America*. Population Trends and Public Policy Series, no. 15. Washington, D.C.: Population Reference Bureau.

O'Neill, O. 1986. *Faces of hunger: An essay on poverty, justice, and development*. London: Allen & Unwin.

O'Riordan, T. 1988. The politics of sustainability. In *Sustainable environmental management: Principles and practice*, ed. R. K. Turner. London: Belhaven Press; Boulder: Westview Press.

Office of Management and Budget. Executive Office of the President of the United States. 1995. *Analytical perspectives: Budget of the United States government, fiscal year 1996*. Washington, D.C.: U.S. Government Printing Office.

Office of Technology Assessment. *See* U.S. Congress. Office of Technology Assessment.

Ohls, J. C., and H. Beebout. 1993. *The food stamp program: Design tradeoffs, policy, and impacts*. Washington, D.C.: The Urban Institute Press.

Organic Farming Research Foundation. n.d. http://www.ofrf.org/giving/index.html. Accessed on 15 December 1999.

Paarlberg, D. 1980. *Farm and food policy issues of the 1980s*. Lincoln: University of Nebraska Press.

Paarlberg, R. 1983. Discussion [of de Janvry, Alain. 1983. Why do governments do what they do? The case of food price policy]. In *The role of markets in the world food economy*, ed. D. G. Johnson and G. E. Schuh. Boulder: Westview Press.

Patel, I. C. Rutgers Cooperative Extension. 1992. Community gardening. New Brunswick: New Jersey Agricultural Experiment Station.

Peck, S. 1989. California farm worker housing. Working Group on Farm Labor and Rural Poverty Working Paper no. 6. Davis: California Institute for Rural Studies.

Pelletier, D. L., V. Kraak, C. McCullum, and U. Uusitalo. 2000. Values, public policy, and community food security. *Agriculture and Human Values* 17:75–93.

Pepper, D. 1993. *Eco-socialism: From deep ecology to social justice*. London: Routledge.

Perez, J. 2002. Community supported agriculture on the central coast. *The Cultivar* (Santa Cruz, Calif., newsletter of the Center for Agroecology and Sustainable Food Systems, University of California, Santa Cruz), 20 (1): 1–3, 18–19.

Perez, J., P. Allen, and M. Brown. 2003. Community supported agriculture on the central coast: The CSA member experience. Research Brief no. 1. Santa Cruz: Center for Agroecology and Sustainable Food Systems, University of California, Santa Cruz.

Peter, G., M. M. Bell, and S. Jarnagin. 2000. Coming back across the fence: Masculinity and the transition to sustainable agriculture. *Rural Sociology* 65 (2): 215–33.

Pfeffer, M. J. 1992. Sustainable agriculture in historical perspective. *Agriculture and Human Values* 9 (4): 4–11.

Pimentel, D. 1993. Soil erosion and agricultural productivity. In *World soil erosion and conservation*, ed. D. Pimentel. Cambridge: Cambridge University Press.

Pimentel, D., and L. Levitan. 1986. Pesticides—amounts applied and amounts reaching pests. *Bioscience* 36 (2): 86–91.

Pimentel, D., E. C. Terhune, R. Dyson-Hudson, S. Rochereau, R. Samis, E. A. Smith, D. Denman, D. Reifschneider, and M. Shepard. 1976. Land degradation: Effects on food and energy resources. *Science* 194 (4261): 149–55.

Pimentel, D., L. McLaughlin, A. Zepp, B. Lakitan, T. Kraus, P. Kleinman, F. Vancini et al. 1991. Environmental and economic impacts of pesticide use. In *Handbook on pest management in agriculture*, ed. D. Pimentel and H. Lehman. Boca Raton, Fla.: LRC Press.

Pisani, D. 1984. *From the family farm to agribusiness: The irrigation crusade in California and the West, 1850–1931*. Berkeley and Los Angeles: University of California Press.

Poincelot, R. P. 1990. From the editor. *Journal of Sustainable Agriculture* 1.

Polanyi, K. 1944. *The great transformation*. Boston: Beacon Press.

Poppendieck, J. 1986. *Breadlines knee-deep in wheat: Food assistance in the Great Depression*. New Brunswick: Rutgers University Press.

———. 1995. Hunger in America: Typification and response. In *Eating agendas: Food and nutrition as social problems*, ed. D. Maurer and J. Sobal. New York: Aldine de Gruyter.

———. 1997. The USA: Hunger in the land of plenty. In *First world hunger: Food security and welfare politics*, ed. G. Riches. London: Macmillan.

Pothukuchi, K., H. Joseph, H. Burton, and A. Fisher. 2002. What's cooking in your food system? A guide to community food assessment. Venice, Calif.: Community Food Security Coalition.

Pothukuchi, K., and J. L. Kaufman. 2000. The food system: A stranger to the planning field. *APA Journal* 66 (2): 113–24.

Pretty, J. N. 1995. *Regenerating agriculture: Policies and practice for sustainability and self-reliance*. London: Earthscan.

———. 1998. The living land: Agriculture, food and community regeneration in the 21st century. London: Earthscan.

Radimer, K. L., C. M. Olson, J. C. Greene, C. C. Campbell, and J. P. Habicht. 1992. Understanding hunger and developing indicators to assess it in women and children. *Journal of Nutrition Education* 24:36S–44S.

Rapoport, A. I. 1998. How has the field mix of academic R&D changed? Division of Science Resource Studies Issue Brief. 2 December. NSF 98-309. Washington, D.C.: National Science Foundation.

Rau, B. 1991. From feast to famine: Official cures and grassroots remedies to Africa's food crisis. London: Zed Press.

Rausser, G. C., and K. R. Farrell, eds. 1985. *Alternative agricultural and food policies and the 1985 farm bill*. Berkeley and Los Angeles: University of California.

Raynolds, L. 2000. Re-embedding global agriculture: The international organic and fair trade movements. *Agriculture and Human Values* 17:297–309.

Redclift, M. 1987. *Sustainable development: Exploring the contradictions.* London: Routledge.
———. 1993. Sustainable development: Concepts, contradictions, and conflicts. In *Food for the future: Conditions and contradictions of sustainability*, ed. P. Allen. New York: John Wiley & Sons.
Reeves, M., A. Katten, and M. Guzman. 2002. Fields of poison 2002: California farmworkers and pesticides. San Francisco: Pesticide Action Network, California Rural Legal Assistance Foundation, and United Farmworkers of America.
Regan, T. 1993. Vegetarianism and sustainable agriculture: The contributions of moral philosophy. In *Food for the future: Conditions and contradictions of sustainability*, ed. P. Allen. New York: John Wiley & Sons.
Riches, G. 1997. Hunger and the welfare state: Comparative perspectives. In *First world hunger: Food security and welfare politics*, ed. G. Riches. New York: St. Martin's Press.
Richmond, E. 1998. Pleas for help flood East Palo Alto hunger program—landlords blamed. *Palo Alto Daily News*, 4 November, 1, 31.
Roberts, R., and G. Hollander. 1997. Sustainable technologies, sustainable farms: Farms, households and structural change. In *Agricultural restructuring and sustainability: A geographical perspective*, ed. B. Ilbery, Q. Chiotti, and T. Rickard. Oxon, U.K.: CAB International.
Robinson, K. L. 1989. *Farm and food policies and their consequences.* Englewood Cliffs, N.J.: Prentice Hall.
Rocheleau, D. E., B. P. Thomas-Slayter, and E. Wangari. 1996. *Feminist political ecology: Global issues and local experience.* London and New York: Routledge.
Rochester, A. 1943. *The populist movement in the United States.* New York: International Publishers.
Rodale, R. 1983. Breaking new ground: The search for a sustainable agriculture. *The Futurist* 1 (1): 15–20.
Rogers, B. L. 1997. Building coalitions for agriculture, nutrition, and the food needs of the poor. In *Visions of American Agriculture*, ed. W. Lockeretz. Ames: Iowa State University Press.
Rosset, P. 1996. Food First and local coalition promote urban farming. *Institute for Food and Development Policy News & Views* 18 (63): 1–2.
Rosset, P. M., and M. A. Altieri. 1997. Agroecology versus input substitution: A fundamental contradiction of sustainable agriculture. *Society & Natural Resources* 10 (3): 283–95.
Roussopoulos, D. I. 1993. *Political ecology: Beyond environmentalism.* Montreal: Black Rose Books.
Ruttan, V. W. 1988. Sustainability is not enough. *American Journal of Alternative Agriculture* 3:128–30.
———, ed. 1992. *Sustainable agriculture and the environment: Perspectives on growth and constraints.* Boulder: Westview Press.
Sachs, C. E. 1991. Women's work and food: A comparative perspective. *Journal of Rural Studies* 7 (1/2): 49–55.
———. 1996. *Gendered fields: Rural women, agriculture, and environment.* Boulder: Westview Press.

Sale, K. 1985. *Dwellers in the land: The bioregional vision*. San Francisco: Sierra Club Books.

SARE. (Sustainable Agriculture Research and Education Program). U.S. Department of Agriculture. 1995. Why SARE works. SARE 1995 project highlights. Washington, D.C.: U.S. Department of Agriculture.

———. 1998a. *SARE 2002—Practical new ideas in agriculture*. Washington, D.C.: U.S. Department of Agriculture.

———. 1998b. The sustainable agriculture research and education (SARE) program. Web site of the Sustainable Agriculture and Research Program. http://www.sarep.ucdavis.edu.

———. 1998c. *Ten years of SARE: A decade of programs, partnership and progress in sustainable agriculture research and education*. Ed. V. Berton. Washington, D.C.: U.S. Department of Agriculture.

Scheie, D. 1997. Creating a more sustainable food and farming system: Lessons from the integrated farming systems initiative. *Consortium News* (newsletter), April–May, 1, 9–10.

Scott, A. 1990. *Ideology and the new social movements*. London: Unwin Hyman.

Scott, J. C. 1989. Everyday forms of resistance. In *Everyday forms of peasant resistance*, ed. F. D. Colburn. Armonk, N.Y.: M. E. Sharpe.

Senate. *See* U.S. Congress. Senate.

Sennett, R., and J. Cobb. 1972. *The hidden injuries of class*. New York: Vintage Books.

Sharp, R. 1995. Organizing for change: People power and the role of institutions. In *The earthscan reader in sustainable development*, ed. J. Kirkby, P. O'Keefe, and L. Timberlake. London: Earthscan.

Shover, J. L. 1976. *First majority—Last minority*. De Kalb: Northern Illinois University Press.

Sims, L. S. 1998. *The politics of fat: Food and nutrition policy in America*. Armonk, N.Y.: M. E. Sharpe.

Slesinger, D. P., and M. J. Pfeffer. 1992. Migrant farm workers. In *Rural poverty in America*, ed. C. M. Duncan. New York: Auburn House.

Smit, J., A. Ratta, and J. Nasr. 1996. *Urban agriculture: Food, jobs and sustainable cities*. Publication Series for Habitat II. New York: United Nations Development Programme.

Smith, N. 1984. *Uneven development: Nature, capital and the production of space*. Oxford: Basil Blackwell.

Somma, M. 1993. Theory building in political ecology. *Social Science Information / Sur les Sciences Sociales* 32 (3): 371–85.

Sommers, P., and J. Smit. 1994. Promoting urban agriculture: A strategy framework for planners in North America, Europe and Asia. Cities Feeding People-Series Report 9. Ottowa: International Development Research Centre. http://www.idrc.ca/cfp/repo9_e.html. Accessed on 15 August 1998.

Soper, K. 1995. *What is nature? Culture, politics, and the non-human*. Oxford: Blackwell.

Speth, J. G. 1992. A new U.S. program for international development and the global environment. *WRI: Issues and Ideas* (Washington, D.C., newsletter of World Resources Institute), March: 1–8.

Stevenson, S., J. Posner, J. Hall, L. Cunningham, and J. Harrison. 1994. Radially organized teams: Addressing the challenges of sustainable agriculture research and extension at land-grant universities. *American Journal of Alternative Agriculture* 9 (1/2): 76–83.

Stiefel, M., and M. Wolfe. 1994. *A voice for the excluded*. London: Zed Books.

Stock, C. M. 1996. *Rural radicals*. New York: Cornell University.

Sustainable Agriculture Network. 1996. *Sustainable agriculture directory of expertise*. 3d ed. Burlington: Sustainable Agriculture Publications.

Swanson, L. E. 1989. The sociology of agriculture: The research agenda. *The Rural Sociologist* 9 (2): 16–26.

Szasz, A. 1994. *Ecopopulism: Toxic waste and the movement for environmental justice*. Minneapolis: University of Minnesota Press.

Thompson, G. D., and J. Kidwell. 1998. Explaining the choice of organic produce: Cosmetic defects, prices, and consumer preferences. *American Journal of Agricultural Economics* 80: 277–87.

Thompson, P. B. 1995. *The spirit of the soil: Agriculture and environmental ethics*. London: Routledge.

Thornborrow, J. 2002. *Power talk: Language and interaction in institutional discourse*. London: Pearson Education.

Thrupp, L. A. 1993. Political ecology of sustainable rural development: Dynamics of social and natural resource degradation. In *Food for the future: Conditions and contradictions of sustainability*, ed. P. Allen. New York: John Wiley & Sons.

Torte, L., and K. Klonsky. 1998. Statistical review of California's organic agriculture 1992–1995. Davis: University of California Agricultural Issues Center.

True, L. 1992. Hunger in the balance: The impact of the proposed AFDC cuts on childhood hunger in California. San Francisco: California Rural Legal Assistance Foundation.

Tuckermanty, E., Director, USDA Community Food Projects Program. 2002. Telephone interview, 8 November.

Turner, R. K. 1988. Sustainability, resource conservation, and pollution control: An overview. In *Sustainable environmental management: Principles and practice*, ed. R. K. Turner. London: Belhaven Press; Boulder: Westview Press.

Tweeten, L. G. 1979. *Foundations of farm policy*. 2d rev. ed. Lincoln: University of Nebraska Press.

UC SAREP. (University of California Sustainable Agriculture Research and Education Program). 1986. *Request for proposals*. Davis: University of California.

———. 1990. Progress report, 1986–1990. Davis: University of California.

———. 1991. *What is sustainable agriculture?* By G. Feenstra, C. Ingels, and D. Campbell. 17 December. Davis: University of California.

———. 1993. Progress report, 1990–1993. Davis: University of California.

———. 2000. Cultivating common ground: Biennial report of the UC Sustainable Agriculture Research and Education Program. Davis: University of California.

———. n.d. What is a community food system? http://www.sarep.ucdavis.edu/cdpp/cfsoverview.htm.

United Farm Workers Union. 1996. The strawberry campaign 1996: Five cents for fairness. Web site of the United Farm Workers. http://www.ufw.org/paper1. htm. Accessed on 19 September 2002.

United Nations World Food Council. 1990. The global state of hunger and malnutrition: 1990 report. Sixteenth Ministerial Session. Bangkok, Thailand: United Nations.

University of Massachusetts Extension. 1997. *What is community supported agriculture and how does it work?* http://www.umass.edu/umext/csa/aboutcsa.html.

U.S. Bureau of the Census. 1987. *Agricultural economics and land ownership survey 1988.* 1987 Census of Agriculture, vol. 3, Related Surveys, part 2. Washington, D.C.: U.S. Government Printing Office.

———. 1996. *A brief look at postwar U.S. income inequality.* By D. H. Weinberg. Current Population Reports—Household Economic Studies. P60-191. Washington, D.C.: U.S. Government Printing Office.

U.S. Conference of Mayors. 1985. *Municipal food policies: How five cities are improving the availability and quality of food for those in need.* Washington, D.C.: U.S. Conference of Mayors.

U.S. Congress. Office of Technology Assessment. 1995a. *Agriculture, trade, and environment: Achieving complementary policies.* May. OTA-ENV-617. Washington, D.C.: U.S. Government Printing Office.

———. 1995b. *Challenges for U.S. agricultural research policy.* September. OTA-ENV-639. Washington, D.C.: U.S. Government Printing Office.

U.S. Congress. House. Committee on Government Operations. 1988. *Low input farming systems: Benefits and barriers.* Seventy-fourth Report by the Committee on Government Operations. 100th Cong., 2d sess., 20 October. House Report 100-1097. Washington, D.C.: U.S. Government Printing Office.

———. Select Committee on Hunger. 1989. *Food security and methods of assessing hunger in the United States.* Serial 101-2. Washington, D.C.: U.S. Government Printing Office.

———. House Committee on Education and Labor. 1991. *Comprehensive Occupational Safety and Health Reform Act, and the fire at the Imperial Food Products Plant in Hamlet, North Carolina.* Statement made by Rep. George Miller (Calif). Hearing on H.R. 3160. 102d Cong., 1st sess., 20 July. Washington, D.C.: U.S. Government Printing Office.

———. Committee on Agriculture. Subcommittee on Department Operations Nutrition and Foreign Agriculture. 1995. *Review of the administration's proposals to reform the food stamp and commodity distribution programs.* 104th Cong., 1st sess., 8 June. Serial No. 104-16. Washington, D.C.: U.S. Government Printing Office.

———. n.d. Members of the Committee on Agriculture: 107th Cong., 2001–2002. Web site of the U.S. House of Representatives. http://agriculture.house.gov. Accessed on 10 January 2003.

U.S. Congress. Senate. Committee on Agriculture, Nutrition, and Forestry. Subcommittee on Nutrition and Investigations. 1990. *Hearing on economic concentration in the meatpacking industry.* 101st Cong., 2d sess., 20 July. Washington, D.C.: U.S. Government Printing Office.

———. 1992. Committee on Agriculture, Nutrition, and Forestry. Subcommittee on Agricultural Research and General Legislation. *Sustainable agriculture: Hearing.* 102d Cong., 2d sess., 17 September. S. Hrg. 102-945. Washington, D.C.: U.S. Government Printing Office.

———. n.d. United States Committee on Agriculture, Nutrition, and Forestry. Web site of the U.S. Senate. http://agriculture.senate.gov. Accessed on 10 January 2003.

U.S. Department of Agriculture. 1980. *Report and recommendations on organic farming.* Washington, D.C.: U.S. Department of Agriculture.

———. 1988. Cooperative State Research Service and Extension Service. *Low-Input/Sustainable Agriculture: Research and Education Program.* Program Brochure. Washington, D.C.: U.S. Department of Agriculture.

———. 1991. *Agriculture and the environment.* Yearbook of Agriculture. Washington, D.C.: U.S. Government Printing Office.

———. 1993a. Cooperative State Research Service. *Dynamics of the research investment: Issues and trends in the agricultural research system.* July. Washington, D.C.: U.S. Department of Agriculture.

———. 1993b. *The Economic Well-Being of Farm Operator Households.* By M. C. Ahearn, J. E. Perry, and H. S. El-Osta. No. 666. Herndon, Va.: U.S. Department of Agriculture.

———. 1994. Economic Research Service. *Economic indicators of the farm sector: National financial summary 1992.* ECIFS 12-1. Herndon, Va.: U.S. Department of Agriculture.

———. 1996. Office of Communications. *Agriculture fact book 1996.* Washington, D.C.: U.S. Government Printing Office.

———. 1997a. Sustainable Agriculture Research and Education Program. *Exploring sustainability in agriculture.* Washington, D.C.: U.S. Department of Agriculture.

———. 1997b. Preliminary summary of food assistance program results for March 1997. May 22, at Washington, D.C.

———. 1998a. Foreign Agricultural Service. *Discussion paper on domestic food security.* 7 April.

———. 1998b. Glickman opens farmers market, announces latest "report card" on the American diet. Web site of the U.S. Department of Agriculture. Accessed on 5 November 1998.

———. 2000. Food and Nutrition Service. *Small Farms/School Meals Initiative.* March. FNS-316. Washington, D.C.: U.S. Department of Agriculture.

———. 2001. Food and agricultural policy: Taking stock for the new century. Washington, D.C.: U.S. Department of Agriculture.

———. 2002. Cooperative State Research Education and Extension Service. *Community Food Projects Competitive Grants Program Request for Applications.* Washington, D.C.: U.S. Department of Agriculture.

———. 2003. Pigford v. Veneman: Consent decree in class action suit by African American farmers. http://www.usda.gov/da/consent.htm. Last updated 9 July 2003. Accessed on 18 August.

———. n.d. Cooperative State Research Education and Extension Service. *Community Food Projects Competitive Grants Program.* Washington, D.C.: U.S. Department of Agriculture.

U.S. Department of Health and Human Services. 1988. *The Surgeon General's report on nutrition and health.* Washington, D.C.: U.S. Government Printing Office.

U.S. Department of Labor. Women's Bureau. 1989. *Facts on working women.* Report 89-4. Washington, D.C.: U.S. Government Printing Office.

USDA. *See* U.S. Department of Agriculture.

U.S. Forest Service. U.S. Department of Agriculture. 1994. *Species endangerment patterns in the United States.* By C. H. Flather, L. A. Joyce, and C. A. Bloomgarden. General Technical Report, RM-241. Ft. Collins, Colo.: Rocky Mountain Forest and Range Experiment Station.

U.S. General Accounting Office. Human Resources Division. 1992a. *Hired farmworkers: Health and well-being at risk.* GAO 92-46. Washington, D.C.: U.S. Government Printing Office.

———. 1992b. *Sustainable agriculture: Program management, accomplishments, and opportunities.* Report to Congressional Requestors. GAO/RCED-92-233. Washington, D.C.: U.S. Government Printing Office.

U.S. Office of Personnel Management. 1992. *Federal civilian workforce statistics: Demographic profile of the federal workforce.* PSO-OWI-5. Springfield, Va.: U.S. Office of Personnel Management.

U.S. Office of the President. 1987. *Economic report of the President.* Transmitted to the Congress January 1987. Washington, D.C.: U.S. Government Printing Office.

U.S. Water Resources Council. 1978. *The nation's water resources: 1975–2000.* 4 vols. Vol. 2, *Second national water assessment.* Washington D.C.: U.S. Government Printing Office.

Van En, R. 1995. Eating for your community. *In Context* 42:29–31.

Villarejo, D. 1990. Environmental effects of living and working in agricultural areas of California: Social and economic factors. In *Health concerns of living and working in agricultural California.* Report of a conference held at the University of California at Davis. Davis, Calif.: Center for Occupational and Environmental Health.

Villarejo, D., D. Lighthall, D. Williams, A. Souter, R. Mines, B. Bade, S. Samuels, and S. McCurdy. 2000. Suffering in silence: A report on the health of California's agricultural workers. Davis: California Institute of Rural Studies.

Vos, T. 2000. Visions of the middle landscape: Organic farming and the politics of nature. *Agriculture and Human Values* 17 (3): 245–56.

W. K. Kellogg Foundation. 1996. Integrated farming systems: Making changes for a strong agricultural future. Battle Creek, Mich.: W. K. Kellogg Foundation.

———. n.d.-a. Effectively engaging farmers and ranchers in food systems change: Lessons learned from Integrated Farming Systems Phase 2. Battle Creek, Mich.: W. K. Kellogg Foundation.

———. n.d.-b. Integrated farming systems overview. http://www.wkkf.org/Programming/Overview.asp. Accessed fall 1998.

Wallerstein, I. 1990. Antisystemic movements: History and dilemmas. In *Transforming the revolution: Social movements and the world-system,* ed. S. Amin, G. Arrigihi, A. G. Frank, and I. Wallerstein. New York: Monthly Review Press.

Wasserstrom, R. F., and R. Wiles. 1985. *Field duty: U.S. farmworkers and pesticide safety.* World Resources Institute Study, no. 3. Washington, D.C.: World Resources Institute.

Watts, M. J., and J. McCarthy. 1997. Nature as artifice, nature as artifact: Development, environment and modernity in the late twentieth century. In *Geographies of economies*, ed. R. Lee and J. Wills. London: Arnold.

Watts, M., and R. Peet. 1996. Conclusion: Towards a theory of liberation ecology. In *Liberation ecologies: Environment, development, and social movements*, ed. R. Peet and M. Watts. London: Routledge.

Weir, M. 1994. Urban poverty and defensive localism. *Dissent* 41 (3): 337–42.

Wells, B. L. 1998. Creating a public space for women in U.S. agriculture: Empowerment, organization, and social change. *Sociologia Ruralis* 38 (3): 371–90.

Western Region SARE. U.S. Department of Agriculture. 1995. Eight years of progress: 1988–1995. Logan, Utah: U.S. Department of Agriculture.

Whatmore, S. 1995. From farming to agribusiness: The global agro-food system. In *Geographies of global change: Remapping the world in the late twentieth century*, ed. R. Johnston, P. Taylor, and M. Watts. Oxford: Blackwell.

Whatmore, S., and L. Thorne. 1997. Nourishing networks: Alternative geographies of food. In *Globalising food: Agrarian questions and global restructuring*, ed. M. Watts. London: Routledge.

Wilkening, E., and J. Gilbert. 1987. Family farming in the United States. In *Family farming in Europe and America*, ed. B. Galeski and E. Wilkening. Boulder: Westview Press.

Williams, R. 1973. Base and superstructure in Marxist cultural theory. *New Left Review* 82 (November–December): 3–16.

———. 1980. *Problems in materialism and culture.* London: Verso.

———. 1981. *The sociology of culture.* Chicago: University of Chicago Press.

Williams-Derry, C., and K. Cook. 2000. Green acre$: How taxpayers are subsidizing the demise of the family farm. Washington, D.C.: Environmental Working Group.

Wimberley, R. C. 1993. Policy perspectives on social, agricultural, and rural sustainability. *Rural Sociology* 58 (1): 1–29.

Winne, M., H. Joseph, and A. Fisher. 1997. Community food security: A guide to concept, design, and implementation. Venice, Calif.: Community Food Security Coalition.

Winson, A. 1993. *The intimate commodity.* Toronto: Garamond Press.

Women Food & Agriculture Network. 1998. Discussion of issues and strategies at agriculture conference. *Newsletter of the Women, Food, & Agriculture Network* 1 (3): 9.

World Bank. 1986. *Poverty and hunger—Issues and options for food security in developing countries.* Washington, D.C.: International Bank for Reconstruction and Development.

———. 2003. *World development indicators 2003.* Washington, D.C.: The World Bank Group.

World Commission on Environment and Development. 1987. *Our common future.* Oxford: Oxford University Press.

Worster, D. 1985. *Rivers of empire*. New York: Oxford University Press.

Wright, A. 1990. The death of Ramon Gonzalez: The modern agricultural dilemma. Austin: University of Texas Press.

Young, D. L. 1989. Policy barriers to sustainable agriculture. *American Journal of Alternative Agriculture* 4 (3 & 4): 135–43.

Young, I. M. 1995. Communication and the Other: Beyond deliberative democracy. In *Justice and identity: Antipodean practices*, ed. M. Wilson and A. Yeatman. Wellington, N.Z.: Bridget Williams Books.

Young, J., ed. 1990. Sustaining the earth: The story of the environmental movement—Its past efforts and future challenges. Cambridge, Mass.: Harvard University Press.

Youngberg, G., N. Schaller, and K. Merrigan. 1993. The sustainable agriculture policy agenda in the United States: Politics and prospects. In *Food for the future: Conditions and contradictions of sustainability*, ed. P. Allen. New York: John Wiley & Sons.